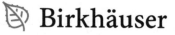

Applied and Numerical Harmonic Analysis

More information about this series at http://www.springer.com/series/4968

Serge Dos Santos • Mostafa Maslouhi
Kasso A. Okoudjou
Editors

Recent Advances in Mathematics and Technology

Proceedings of the First International
Conference on Technology, Engineering,
and Mathematics, Kenitra, Morocco,
March 26-27, 2018

 Birkhäuser

Editors

Serge Dos Santos
Institut National des Sciences Appliquées
Centre Val de Loire
Blois, France

Mostafa Maslouhi
Informatics, Logistics and Mathematics
Université Ibn-Tofail
Kenitra, Morocco

Kasso A. Okoudjou
Norbert Wiener Center
Department of Mathematics
University of Maryland
College Park, MD, USA

ISSN 2296-5009   ISSN 2296-5017   (electronic)
Applied and Numerical Harmonic Analysis
ISBN 978-3-030-35204-2   ISBN 978-3-030-35202-8   (eBook)
https://doi.org/10.1007/978-3-030-35202-8

Mathematics Subject Classification: 42B10, 37Mxx, 82-06

This book is published under the imprint Birkhäuser, www.birkhauser-science.com by the registered company Springer Nature Switzerland AG.
The registered company address is: Gewerbestrasse 11, 6330 Cham, Switzerland

*This book is dedicated to the authors'
children:
Youssef, Mariam, and Mouad Maslouhi;
Shadeh, Shola, and Femi Okoudjou;
Lauryne, Maëlys, Lazare, and Alcide Dos
Santos.*

ANHA Series Preface

The *Applied and Numerical Harmonic Analysis (ANHA)* book series aims to provide the engineering, mathematical, and scientific communities with significant developments in harmonic analysis, ranging from abstract harmonic analysis to basic applications. The title of the series reflects the importance of applications and numerical implementation, but richness and relevance of applications and implementation depend fundamentally on the structure and depth of theoretical underpinnings. Thus, from our point of view, the interleaving of theory and applications and their creative symbiotic evolution is axiomatic.

Harmonic analysis is a wellspring of ideas and applicability that has flourished, developed, and deepened over time within many disciplines and by means of creative cross-fertilization with diverse areas. The intricate and fundamental relationship between harmonic analysis and fields such as signal processing, partial differential equations (PDEs), and image processing is reflected in our state-of-the-art *ANHA* series.

Our vision of modern harmonic analysis includes mathematical areas such as wavelet theory, Banach algebras, classical Fourier analysis, time-frequency analysis, and fractal geometry, as well as the diverse topics that impinge on them.

For example, wavelet theory can be considered an appropriate tool to deal with some basic problems in digital signal processing, speech and image processing, geophysics, pattern recognition, biomedical engineering, and turbulence. These areas implement the latest technology from sampling methods on surfaces to fast algorithms and computer vision methods. The underlying mathematics of wavelet theory depends not only on classical Fourier analysis, but also on ideas from abstract harmonic analysis, including von Neumann algebras and the affine group. This leads to a study of the Heisenberg group and its relationship to Gabor systems, and of the metaplectic group for a meaningful interaction of signal decomposition methods. The unifying influence of wavelet theory in the aforementioned topics illustrates the justification for providing a means for centralizing and disseminating information from the broader, but still focused, area of harmonic analysis. This will be a key role of *ANHA*. We intend to publish with the scope and interaction that such a host of issues demands.

Along with our commitment to publish mathematically significant works at the frontiers of harmonic analysis, we have a comparably strong commitment to publish major advances in the following applicable topics in which harmonic analysis plays a substantial role:

Antenna theory   *Prediction theory*
Biomedical signal processing   *Radar applications*
Digital signal processing   *Sampling theory*
Fast algorithms   *Spectral estimation*
Gabor theory and applications   *Speech processing*
Image processing   *Time-frequency and*
Numerical partial differential equations   *time-scale analysis*
                                                          Wavelet theory

The above point of view for the *ANHA* book series is inspired by the history of Fourier analysis itself, whose tentacles reach into so many fields.

In the last two centuries Fourier analysis has had a major impact on the development of mathematics, on the understanding of many engineering and scientific phenomena, and on the solution of some of the most important problems in mathematics and the sciences. Historically, Fourier series were developed in the analysis of some of the classical PDEs of mathematical physics; these series were used to solve such equations. In order to understand Fourier series and the kinds of solutions they could represent, some of the most basic notions of analysis were defined, e.g., the concept of "function." Since the coefficients of Fourier series are integrals, it is no surprise that Riemann integrals were conceived to deal with uniqueness properties of trigonometric series. Cantor's set theory was also developed because of such uniqueness questions.

A basic problem in Fourier analysis is to show how complicated phenomena, such as sound waves, can be described in terms of elementary harmonics. There are two aspects of this problem: first, to find, or even define properly, the harmonics or spectrum of a given phenomenon, e.g., the spectroscopy problem in optics; second, to determine which phenomena can be constructed from given classes of harmonics, as done, for example, by the mechanical synthesizers in tidal analysis.

Fourier analysis is also the natural setting for many other problems in engineering, mathematics, and the sciences. For example, Wiener's Tauberian theorem in Fourier analysis not only characterizes the behavior of the prime numbers, but also provides the proper notion of spectrum for phenomena such as white light; this latter process leads to the Fourier analysis associated with correlation functions in filtering and prediction problems, and these problems, in turn, deal naturally with Hardy spaces in the theory of complex variables.

Nowadays, some of the theory of PDEs has given way to the study of Fourier integral operators. Problems in antenna theory are studied in terms of unimodular trigonometric polynomials. Applications of Fourier analysis abound in signal processing, whether with the fast Fourier transform (FFT), or filter design, or the adaptive modeling inherent in time-frequency-scale methods such as wavelet theory. The coherent states of mathematical physics are translated and modulated Fourier transforms, and these are used, in conjunction with the uncertainty principle, for dealing with signal reconstruction in communications theory. We are back to the raison d'être of the *ANHA* series!

University of Maryland   John J. Benedetto
College Park   Series Editor

Preface

One of the impacts of the trifecta Technology, Engineering, and Mathematics (TEM) on our daily life is the enormous amount of data we generate. For example, Technology and Engineering are increasingly becoming the main source of "big data" production. To analyze, organize, process, and act upon these data, researchers in both academia and industry are devising new paradigms. These include powerful machine learning algorithms, especially deep learning models such as convolutional neural networks (CNNs), which have recently achieved outstanding predictive performance in a wide range of multimedia applications, including visual object classification, scene understanding, speech recognition, and activity prediction. Many of these new applications are generally based on advances in mathematics, and particularly, mathematical modeling, optimization, numerical analysis and simulations, mathematical signal processing, and computer sciences.

From the modern engineering point of view, the forthcoming digital transformation also triggers opportunities in new and growing fields including big data analytics, artificial intelligence, automation, and imaging. Advances in digital technologies for industries like augmented reality, virtual reality, and mixed reality will also see valuable changes and developments in education and training delivery using modern mathematics.

To offer a forum for researchers working in these fields to discuss the advances and the challenges created by these new paradigms, the first edition of the Technology, Engineering and Mathematics Conference (TEM18) was organized, March 26 and 27, 2018, in Kenitra, Morocco (see http://ensa.uit.ac.ma/tem2018/ for more details). It brought together a group of renowned researchers and professionals both from academia and industries who presented their work on topics that were thematically divided as follows:

- Big Data Analytics and Applications
- Biomathematics
- Computer Engineering and Applications
- Economics and Financial Engineering
- Harmonic Analysis

- Medical Imaging and Non-Destructive Testing
- Numerical Analysis and Modeling
- Optimization and Control
- Smart Technologies and Engineering
- Stochastic and Statistics

This volume grew out of the conference and includes papers based on some of the presentations. It is designed to broadly engage the "Computer Sciences and Smart Technologies" and the "Mathematical Modeling" communities on the problems and challenges presented by the aforementioned paradigms.

Keeping with the theme of the conference, the volume is divided into the following three parts:

- Part I entitled *Mathematical modeling* contains three papers/chapters centered around harmonic analysis and differential equations both deterministic and stochastic.
- Part II entitled *Advanced mathematics for imaging* is made up of three papers/chapters dealing with image processing in general, and applications to Non-Destructive Testing (NDT) and medical imaging (US and MRI) in particular.
- Part III entitled *Computer Sciences and Smart Technologies* assembles four papers/chapters dealing with applications in smart technologies such as improvement of network lifetime or improvement of e-learning processes with fuzzy logic and artificial intelligence.

To close, we would like to thank Professors Mohammed El Fatini and Hanaa El Hachimi, both members of the organizing committee, for their help putting together a successful conference. We also acknowledge the generous support of Professor Nabil Hmina, Director of the National School of Applied Sciences, and Professor Azeddine EL Midaoui, President of Ibn Tofail University Morocco.

Blois, France   Serge Dos Santos
Kenitra, Morocco   Mostafa Maslouhi
College Park, MD, USA   Kasso A. Okoudjou
April 2019

Contents

Part II Advanced Mathematics for Imaging

Contributors

Otman Abdoun Laboratory of Advanced Sciences and Technologies, Polydisciplinary Faculty, Abdelmalek Essaadi University, Larache, Morocco

Khadija Akdim Department of Mathematics, Faculty of Sciences and Techniques, Cadi Ayyad University, Marrakech, Morocco

Ilham Amezzane LaRIT Lab, Faculty of Sciences, Ibn Tofail University, Kenitra, Morocco

Aouatif Amine BOSS Team, LGS Laboratory, ENSA of Kenitra, Ibn Tofail University, Kenitra, Morocco

Moulay Driss Aouragh MAMCS Group, Department of Maths, M2I Laboratory, FST, Moulay Ismaïl University, Errachidia, Morocco

Mohamed Bakhouya LERMA Lab, Faculty of Computing and Logistics, International University of Rabat, Sala Aljadida, Morocco

Thomas Deregnaucourt CNRS-UCA UMR 5961, Clermont-Ferrand, France

Serge Dos Santos Institut National des Sciences Appliquées, Centre Val de Loire, Blois, France

Mohamed El Aroussi LaRIT Lab, Faculty of Sciences, Ibn Tofail University, Kenitra, Morocco

Hanan El Bhilat National Higher School of Electricity and Mechanics, Laboratory of Control and Mechanical Characterization of Materials and Structures, Casablanca, Morocco

Abderrahman El Boukili Department of Physics, FST, Moulay Ismaïl University, Errachidia, Morocco

Khalid El Had Higher Institute of Maritime Studies, Laboratory of Mechanics, Casablanca, Morocco

Mohammed El Kassimi Department of Mathematics and Computer Sciences, Faculty of Sciences-Meknès, Equipe d'Analyse Harmonique et Probabilités, Moulay Ismaïl University, Meknès, Morocco

Abdelmoujib Elkhoumri Ibn Tofail University, Kenitra, Morocco

Saïd Fahlaoui Department of Mathematics and Computer Sciences, Faculty of Sciences-Meknès, Equipe d'Analyse Harmonique et Probabilités, Moulay Ismaïl University, Meknès, Morocco

Youssef Fakhri LaRIT Lab, Faculty of Sciences, Ibn Tofail University, Kenitra, Morocco

Abdelilah Hachim Higher Institute of Maritime Studies, Laboratory of Mechanics, Casablanca, Morocco

El Khatir Haimoudi Laboratory of Advanced Sciences and Technologies, Polydisciplinary Faculty, Abdelmalek Essaadi University, Larache, Morocco

Fatima-Zohra Hibbi Laboratory of Advanced Sciences and Technologies, Polydisciplinary Faculty, Abdelmalek Essaadi University, Larache, Morocco

Jalal Laassiri Ibn Tofail University, Kenitra, Morocco

Abdelmajid Oumnad Research Team in Smart Communications-ERSC-Research Centre E3S, EMI, Mohamed V University, Rabat, Morocco

Amine Rghioui Research Team in Smart Communications-ERSC-Research Centre E3S, EMI, Mohamed V University, Rabat, Morocco

Zakaria Sabir BOSS Team, LGS Laboratory, ENSA of Kenitra, Ibn Tofail University, Kenitra, Morocco

Houda Salmi National Higher School of Electricity and Mechanics, Laboratory of Control and Mechanical Characterization of Materials and Structures, Casablanca, Morocco

Chafik Samir CNRS-UCA UMR 5961, Clermont-Ferrand, France

Acronyms

AH	Adaptive hypermedia
AI	Artificial intelligence
BD	Big data
BSDE	Backward stochastic differential equation
CCN	Content centric networking
CPU	Central processing unit
CS	Content store
CT	Compact tension
CUDA	Compute Unified Device Architecture
DNA	Deoxyribonucleic acid
FEM	Finite element method
FIA	Future internet architectures
FIB	Forwarding information base
FIS	Fuzzy inference systems
FPGA	Field programmable gate array
FS	Future selection
GP	Gaussian Process
GPU	Graphics Processing Unit
GQFT	Gabor quaternionic Fourier transform
GSM	Global system for mobile
HAR	Human activity recognition
HDFS	Hadoop distributed file system
IBK	K-nearest neighbor
IMQ	Inverse multi-quadratic
IoT	Internet of Things
ITS	Intelligent tutoring system
LDDMM	Large deformation diffeomorphic metric mapping
LEFM	Linear elastic fracture mechanics
ML	Machine learning
MLP	Multilevel perceptron

MQ	Multi-quadratic
MRI	Magnetic resonance imaging
MRS	Multi-platform for road safety
MST	Model of smart tutoring
NDN	Named Data Networking
NDT	Non-destructive testing
NDT 4.0	NDT within the industrial process Industry 4.0
NEWS	Nonlinear elastic wave spectroscopy
NIT	Non-invasive testing
NS	Network simulator
OHD-SVM	Optimized hierarchical decomposition SVM
OVA	One-versus-all
PI	Pulse inversion
PIT	Pending Interest Table
PLL	Phase-locked loop
PRC	Precision-recall curve
PUFEM	Partition of the Unit Finite Element Method
QFT	Quaternion Fourier transform
QP	Quadratic programming
RBF	Radial basis function
RBSDE	Reflected backward stochastic differential equation
RMSE	Root Mean Square Error
ROC	Receiver operating characteristic
RSU	Roadside unit
RTT	Round trip time
SDE	Stochastic differential equation
SDR	Stochasticity to determinism ratio
SDS	Secure Data Services
SEC	Stimulus evaluation checks
SIF	Stress intensity factor
SMO	Sequential Minimal Optimization
SMS	Short message service
SNR	Signal-to-noise ratio
SRV	Square Root Velocity
STS	Smart tutoring system
SVM	Support vector machines
TCP/IP	Transmission Control Protocol/Internet Protocol
TPS	Thin plate splines
TR	Time reversal

TR-NEWS	Time reversal based nonlinear elastic wave spectroscopy
UCA	Ultrasound contrast agents
US	Ultrasound
WBAN	Wireless Body Area Network
WFT	Wavelet Fourier transform
WRR	Weighted Round-Robin
XFEM	Extended finite element method

Part I
Mathematical Modeling

In this first part, we present the papers that are more mathematical in nature. In particular, the first chapter written by M. El Kassim and S. Fahlaoui introduces a two-sided Gabor Quaternic Fourier transform and develops some related uncertainty principles. In Chap. 2, K. Akdim investigates the existence and the uniqueness of solution for a family of reflected backward stochastic differential equations in a convex polyhedron. Finally, in the last chapter of this part, M. D. Aouragh and A. El Boukili analyze the decay of energy of a non-homogeneous hybrid systems of elasticity.

Chapter 1
The Two-Sided Gabor Quaternionic Fourier Transform and Uncertainty Principles

Mohammed El Kassimi and Saïd Fahlaoui

Abstract In this paper, we define a new transform called the Gabor quaternionic Fourier transform (GQFT), which generalizes the classical windowed Fourier transform to quaternion-valued signals, and we give several important properties such as the Plancherel formula and inversion formula. Finally, we establish the Heisenberg uncertainty principles for the GQFT.

Keywords Quaternion algebra · Quaternionic transform · Gabor transform · Heisenberg uncertainty · Logarithmic uncertainty

1.1 Introduction

As it is known, the quaternion Fourier transform (QFT) is a very useful mathematical tool. It has been discussed extensively in the literature and has proved to be powerful and useful in some theories. In [1, 4, 7] the authors used the (QFT) to extend the color image analysis. Researchers in [3] applied the QFT to image processing and neural computing techniques. The QFT is a generalization of the real and complex Fourier transform (FT), but it is ineffective in representing and computing local information about quaternionic signals. A lot of papers have been devoted to the extension of the theory of the windowed FT to the quaternionic case. Recently Bülow and Sommer [4, 5] extend the WFT to the quaternion algebra. They introduced a special case of the GQFT known as quaternionic Gabor filters. They applied these filters to obtain a local two-dimensional quaternionic phase. In [2] Bahri et al. studied the right-sided windowed quaternion Fourier transform.

M. El Kassimi (✉) · S. Fahlaoui
Department of Mathematics and Computer Sciences, Faculty of Sciences-Meknès, Equipe d'Analyse Harmonique et Probabilités, Moulay Ismaïl University, Meknès, Morocco
e-mail: m.elkassimi@edu.umi.ac.ma; s.fahlaoui@fs.umi.ac.ma

© Springer Nature Switzerland AG 2020
S. Dos Santos et al. (eds.), *Recent Advances in Mathematics and Technology*,
Applied and Numerical Harmonic Analysis,
https://doi.org/10.1007/978-3-030-35202-8_1

In [9] the authors studied two-sided windowed (QFT) for the case when the window has a real value. In this paper, we study the two-sided quaternionic Gabor Fourier transform (GQFT) with the window quaternionic value and some important properties are derived. We start by reminding some results of two-sided quaternionic Fourier transform (QFT), we give some examples to show the difference between the GQFT and WFT, and we establish important properties of the GQFT like inversion formula, Plancherel formula, using a version of Heisenberg uncertainty principle for two-sided QFT to prove a generalized uncertainty principle for GQFT.

1.2   Definition and Properties of Quaternion $\mathbb{H}$

The quaternion algebra $\mathbb{H}$ is defined over $\mathbb{R}$ with three imaginary units i, j, and k that obey the Hamilton's multiplication rules,

$$ij = -ji = k, \quad jk = -kj = i, \quad ki = -ik = j \tag{1.1}$$

$$i^2 = j^2 = k^2 = ijk = -1 \tag{1.2}$$

According to (1.1) $\mathbb{H}$ is non-commutative, and one cannot directly extend various results on complex numbers to a quaternion. For simplicity, we express a quaternion q as the sum of scalar q_1, and a pure 3D quaternion q. Every quaternion can be written explicitly as

$$q = q_1 + iq_2 + jq_3 + kq_4 \in \mathbb{H}, \quad q_1, q_2, q_3, q_4 \in \mathbb{R}$$

The conjugate of quaternion q is obtained by changing the sign of the pure part, i.e.

$$\overline{q} = q_1 - iq_2 - jq_3 - kq_4$$

The quaternion conjugation is a linear anti-involution

$$\overline{\overline{p}} = p, \quad \overline{p+q} = \overline{p} + \overline{q}, \quad \overline{pq} = \overline{q}\,\overline{p}, \quad \forall p, q \in \mathbb{H}$$

The modulus $|q|$ of a quaternion q is defined as

$$|q| = \sqrt{q\overline{q}} = \sqrt{q_1^2 + q_2^2 + q_3^2 + q_4^2}, \quad |pq| = |p||q|.$$

It is straightforward to see that

$$|pq| = |p||q|, |q| = |\overline{q}|, p, q \in \mathbb{H}$$

In particular, when $q = q_1$ is a real number, the module $|q|$ reduces to the ordinary Euclidean modulus, i.e., $|q| = \sqrt{q_1 q_1}$. A function $f : \mathbb{R}^2 \to \mathbb{H}$ can be expressed as

$$f(x, y) := f_1(x, y) + i f_2(x, y) + j f_3(x, y) + k f_4(x, y)$$

where $(x, y) \in \mathbb{R} \times \mathbb{R}$.

We introduce an inner product of functions f, g defined on $\mathbb{R}^2$ with values in $\mathbb{H}$ as follows:

$$< f, g >_{L^2(\mathbb{R}^2, \mathbb{H})} = \int_{\mathbb{R}^2} f(x)\overline{g(x)}dx$$

If $f = g$ we obtain the associated norm by

$$\|f\|_2^2 = < f, f >_2 = \int_{\mathbb{R}^2} |f(x)|^2 dx$$

The space $L^2(\mathbb{R}^2, \mathbb{H})$ is then defined as

$$L^2(\mathbb{R}^2, \mathbb{H}) = \{f | f : \mathbb{R}^2 \to \mathbb{H}, \|f\|_2 < \infty\}$$

And we define the norm of $L^2(\mathbb{R}^2, \mathbb{H})$ by

$$\|f\|_{L^2(\mathbb{R}^2, \mathbb{H})}^2 = \|f\|_2^2$$

1.3 The Two-Sided Gabor Quaternionic Fourier Transform (GQFT)

The quaternion Fourier transform (QFT) is an extension of Fourier transform proposed by Ell [8]. Due to the non-commutative properties of quaternion, there are three different types of QFT: the left-sided QFT, the right-sided QFT, and the two-sided QFT [12]. In this paper we only treat the two-sided QFT. We now review the definition and some properties of the two-sided QFT[11].

Definition 1.1 (Quaternion Fourier Transform) The two-sided quaternion Fourier transform (QFT) of a quaternion function $f \in L^1(\mathbb{R}^2, \mathbb{H})$ is the function $\mathcal{F}_q(f) : \mathbb{R}^2 \to \mathbb{H}$ defined by
for $\omega = (\omega_1, \omega_2) \in \mathbb{R} \times \mathbb{R}$

$$\mathcal{F}_q(f)(w) = \int_{\mathbb{R}^2} e^{-2\pi i x_1.\omega_1} f(x) e^{-2\pi j x_2.\omega_2} dx \tag{1.3}$$

where $dx = dx_1 dx_2$

This transform can be inverted by means of

Theorem 1.1 *If $f, \mathcal{F}_q(f) \in L^2(\mathbb{R}^2, \mathbb{H})$, then,*

$$f(x) = \mathcal{F}_q^{-1}\mathcal{F}_q(f)(x) = \int_{\mathbb{R}^2} e^{2\pi i x_1 \cdot \omega_1} \mathcal{F}_q(f)(\omega) e^{2\pi j x_2 \cdot \omega_2} d\omega \qquad (1.4)$$

Theorem 1.2 (Plancherel Theorem for QFT) *If $f \in L^2(\mathbb{R}^2, \mathbb{H})$ then*

$$\|f\|_2 = \|\mathcal{F}_q(f)\|_2 \qquad (1.5)$$

Proof See [11]. □

Definition 1.2 A quaternion window function is a non-null function $\varphi \in L^2(\mathbb{R}^2, \mathbb{H})$

Based on the above formula (1.3) for the QFT, we establish the following definition of the two-sided Gabor quaternionic Fourier transform (GQFT).

Definition 1.3 We define the GQFT of $f \in L^2(\mathbb{R}^2, \mathbb{H})$ with respect to non-zero quaternion window function $\varphi \in L^2(\mathbb{R}^2, \mathbb{H})$ as,

$$G_\varphi f(\omega, b) = \int_{\mathbb{R}^2} e^{-2\pi i x_1 \omega_1} f(x)\overline{\varphi(x - b)} e^{-2\pi j x_2 \cdot \omega_2} dx \qquad (1.6)$$

Note that the order of the exponentials in (1.6) is fixed because of the non-commutativity of the product of quaternion.

The energy density is defined as the modulus square of GQFT (Definition 1.3) given by

$$|G_\varphi f(\omega, b)|^2 = |\int_{\mathbb{R}^2} e^{-2i\pi x_1 \omega_1} f(x)\overline{\varphi(x - b)} e^{-2j\pi x_2 \omega_2} dx|^2 \qquad (1.7)$$

Equation (1.7) is often called a spectrogram which measures the energy of a quaternion-valued function f in the position-frequency neighborhood of (ω, b).

1.3.1 Examples of the GQFT

For illustrative purposes, we shall discuss examples of the GQFT. We begin with a straightforward example.

Example 1.1 Consider the two-dimensional window function defined by

$$\varphi(x) = \begin{cases} 1, & \text{for } -1 \le x_1 \le 1 \text{ and } -1 \le x_2 \le 1; \\ 0, & \text{otherwise} \end{cases} \qquad (1.8)$$

$$f(x) = \begin{cases} e^{-x_1 - x_2}, 0 \le x_1 \le +\infty \text{ and } 0 \le x_2 \le +\infty; \\ 0, & \text{otherwise} \end{cases} \qquad (1.9)$$

we start by explaining that we have $b = (b_1, b_2) \in \mathbb{R}^2$, then,

$$\varphi(x - b) = \begin{cases} 1, & \text{for } -1 \le x_1 - b_1 \le 1 \text{ and } -1 \le x_2 - b_2 \le 1; \\ 0, & \text{otherwise} \end{cases}$$

from which we give

$$\varphi(x - b) = \begin{cases} 1, & \text{for } -1 + b_1 \le x_1 \le 1 + b_1 \text{ and } -1 + b_2 \le x_2 \le 1 + b_2; \\ 0, & \text{otherwise} \end{cases}$$

By applying the definition of the GQFT we have

$$\mathcal{G}_\varphi f(\omega, b) = \int_{\mathbb{R}^2} e^{-i2\pi x_1 \omega_1} f(x)\overline{\varphi(x - b)} e^{-j2\pi x_2 \omega_2} dx,$$

$$= \int_{m_1}^{1+b_1} \int_{m_2}^{1+b_2} e^{-i2\pi x_1 \omega_1} e^{-x_1 - x_2} e^{-j2\pi x_2 \omega_2} dx_1 dx_2,$$

with $m_1 = max(0, -1 + b_1)$; $m_2 = max(0, -1 + b_2)$,

$$= \int_{m_1}^{1+b_1} e^{-x_1(1 + i2\pi\omega_1)} dx_1 \int_{m_2}^{1+b_2} e^{-x_2(1 + j2\pi\omega_2)} dx_2,$$

$$= \left[\frac{e^{-x_1(1 + i2\pi\omega_1)}}{(-1 - i2\pi\omega_1)} \right]_{m_1}^{1+b_1} \left[\frac{e^{-x_2(1 + j2\pi\omega_2)}}{(-1 - j2\pi\omega_2)} \right]_{m_2}^{1+b_2},$$

$$= \frac{1}{(-1 - i2\pi\omega_1)(-1 - j2\pi\omega_2)} \left(e^{-(1+b_1)(1 + i2\pi\omega_1)} - e^{-m_1(1 + i2\pi\omega_1)} \right)$$
$$\left(e^{-(1+b_2)(1 + j2\pi\omega_2)} - e^{-m_2(1 + j2\pi\omega_2)} \right)$$

Example 1.2 Given the window function of the two-dimensional Haar function defined by

$$\varphi(x) = \begin{cases} 1, & \text{for } 0 \le x_1 \le \frac{1}{2} \text{ and } 0 \le x_2 \le \frac{1}{2}; \\ -1, & \text{for } \frac{1}{2} \le x_1 \le 1 \text{ and } \frac{1}{2} \le x_2 \le 1; \\ 0, & \text{otherwise} \end{cases} \tag{1.10}$$

find the GQFT of the Gaussian function $f(x) = e^{-(x_1^2 + x_2^2)}$.

By the definition of the function φ in (1.10), we have

$$\varphi(x - b) = \begin{cases} 1, & \text{for } b_1 \le x_1 \le \frac{1}{2} + b_1 \text{ and } b_2 \le x_2 + b_2 \le b_2 + \frac{1}{2}; \\ -1, & \text{for } \frac{1}{2} + b_1 \le x_1 \le 1 + b_1 \text{ and } \frac{1}{2} + b_2 \le x_2 \le 1 + b_2; \\ 0, & \text{otherwise} \end{cases}$$

From Definition 1.3 and by the separation of the variables, we obtain

$$\mathcal{G}_\varphi\{f\}(\omega, b) = \int_{\mathbb{R}^2} e^{-i2\pi x_1 \omega_1} f(x)\overline{\varphi(x - b)} e^{-j2\pi x_2 \omega_2} dx,$$

$$= \int_{b_1}^{\frac{1}{2}+b_1} e^{-i2\pi x_1 \omega_1} e^{-x_1^2} dx_1 \int_{b_2}^{\frac{1}{2}+b_2} e^{-x_2^2} e^{-j2\pi x_2 \omega_2} dx_2,$$

$$- \int_{\frac{1}{2}+b_1}^{1+b_1} e^{-i2\pi x_1 \omega_1} e^{-x_1^2} dx_1 \int_{\frac{1}{2}+b_2}^{1+b_2} e^{-x_2^2} e^{-j2\pi x_2 \omega_2} dx_2$$

by completing squares, we have

$$\mathcal{G}_\varphi\{f\}(\omega, b) = \int_{b_1}^{\frac{1}{2}+b_1} e^{-(x_1 + i\pi\omega_1)^2} e^{-(\omega_1 \pi)^2} dx_1 \int_{b_2}^{\frac{1}{2}+b_2} e^{-(x_2 + j\pi\omega_2)^2} e^{-(\omega_2 \pi)^2} dx_2$$

$$- \int_{\frac{1}{2}+b_1}^{1+b_1} e^{-(x_1 + i\pi\omega_1)^2} e^{-(\omega_1 \pi)^2} dx_1 \int_{\frac{1}{2}+b_2}^{1+b_2} e^{-(x_2 + j\pi\omega_2)^2} e^{-(\omega_2 \pi)^2} dx_2$$

we factorize and we get

$$\mathcal{G}_\varphi\{f\}(\omega, b) = e^{-\omega_1^2 \pi^2} \left(\int_{b_1}^{\frac{1}{2}+b_1} e^{-(x_1 + i\pi\omega_1)^2} dx_1 \right) e^{-\omega_2^2 \pi^2} \left(\int_{b_2}^{\frac{1}{2}+b_2} e^{-(x_2 + j\pi\omega_2)^2} dx_2 \right)$$

$$- e^{-\omega_1^2 \pi^2} \left(\int_{\frac{1}{2}+b_1}^{1+b_1} e^{-(x_1 + i\pi\omega_1)^2} dx_1 \right) e^{-\omega_2^2 \pi^2} \left(\int_{\frac{1}{2}+b_2}^{1+b_2} e^{-(x_2 + j\pi\omega_2)^2} dx_2 \right)$$

making the substitutions $y_1 = x_1 + i\pi\omega_1$ and $y_2 = x_2 + j\pi\omega_2$ in the above expression we immediately obtain

$$
\mathcal{G}_\varphi\{f\}(\omega, b) = e^{-\omega_1^2\pi^2}\left(\int_{b_1+i\pi\omega_1}^{\frac{1}{2}+b_1+i\pi\omega_1} e^{-y_1^2}dy_1\right)e^{-\omega_2^2\pi^2}\left(\int_{b_2+j\pi\omega_2}^{\frac{1}{2}+b_2+j\pi\omega_2} e^{-y_2^2}dy_2\right)
$$
$$
- e^{-\omega_1^2\pi^2}\left(\int_{\frac{1}{2}+b_1+i\pi\omega_1}^{1+b_1+i\pi\omega_1} e^{-y_1^2}dy_1\right)e^{-\omega_2^2\pi^2}\left(\int_{\frac{1}{2}+b_2+j\pi\omega_2}^{1+b_2+j\pi\omega_2} e^{-y_2^2}dy_2,\right)
$$
$$
= e^{-\omega_1^2\pi^2}\left(\int_0^{b_1+i\pi\omega_1}(-e^{-y_1^2})dy_1 + \int_0^{\frac{1}{2}+b_1+i\pi\omega_1} e^{-y_1^2}dy_1\right)
$$
$$
\times\, e^{-\omega_2^2\pi^2}\left(\int_0^{b_2+j\pi\omega_2}(-e^{-y_2^2})dy_2 + \int_0^{\frac{1}{2}+b_2+j\pi\omega_2} e^{-y_2^2}dy_2\right)
$$
$$
- e^{-\omega_1^2\pi^2}\left(\int_0^{\frac{1}{2}+b_1+i\pi\omega_1}(-e^{-y_1^2})dy_1 + \int_0^{1+b_1+i\pi\omega_1} e^{-y_1^2}dy_1\right)
$$
$$
\times\, e^{-\omega_2^2\pi^2}\left(\int_0^{\frac{1}{2}+b_2+j\pi\omega_2}(-e^{-y_2^2})dy_2 + \int_0^{1+b_2+j\pi\omega_2} e^{-y_2^2}dy_2\right) \quad (1.11)
$$

we can write the Eq. (1.11) in the form

$$
\mathcal{G}_\varphi\{f\}(\omega, b) = e^{-\omega_1^2\pi^2}\left[-qf(b_1 + i\pi\omega_1) + qf(\frac{1}{2} + b_1 + i\pi\omega_1)\right]
$$
$$
\times\, e^{-\omega_2^2\pi^2}\left[-qf(b_2 + j\pi\omega_2) + qf(\frac{1}{2} + b_2 + j\pi\omega_2)\right]
$$
$$
- e^{-\omega_1^2\pi^2}\left[-qf(\frac{1}{2} + b_1 + i\pi\omega_1) + qf(1 + b_1 + i\pi\omega_1)\right]
$$
$$
\times\, e^{-\omega_2^2\pi^2}\left[-qf(\frac{1}{2} + b_2 + j\pi\omega_2) + qf(1 + b_2 + j\pi\omega_2)\right]
$$

where, $qf(x) = \int_0^x e^{-t^2}dt$

1.4   Properties of GQFT

In this section, we are going to give some properties for the Gabor quaternionic Fourier transform.

Theorem 1.3 *Let $f \in L^2(\mathbb{R}^2, \mathbb{H})$; and $\varphi \in L^2(\mathbb{R}^2, \mathbb{H})$ be a non-zero quaternionic window function. Then, we have*

$$(\mathcal{G}_\varphi\{T_y f\}(\omega, b) = e^{-2i\pi y_1\omega_1}(\mathcal{G}_\varphi f)(\omega, b - y)e^{-2j\pi x_2.\omega_2} \qquad (1.12)$$

where $T_y f(x) = f(x - y)$; and $y = (y_1, y_2) \in \mathbb{R}^2$

Proof We have

$$\mathcal{G}_\varphi\{T_y f\}(w, b) = \int_{\mathbb{R}^2} e^{-2\pi i x_1\omega_1} f(x)\overline{\varphi(x - b)}e^{-2\pi j x_2.\omega_2}dx$$

we take $t = x - y$, then

$$\mathcal{G}_\varphi\{T_y f\}(w, b) = \int_{\mathbb{R}^2} e^{-2i\pi(t_1+y_1)\omega_1} f(x)\overline{\varphi(t + y - b)}e^{-2j\pi(t_2+y_2)\omega_2}dt$$

$$= e^{-2i\pi y_1\omega_1}\int_{\mathbb{R}^2} e^{-2i\pi t_1\omega_1} f(x)\overline{\varphi(t + y - b)}e^{-2j\pi t_2\omega_2}dt \ e^{-2i\pi y_2\omega_2}$$

$$= e^{-2i\pi y_1\omega_1}\mathcal{G}_\varphi\{f\}(\omega, b - y)e^{-2j\pi x_2.\omega_2}$$

$$\square$$

Theorem 1.4 *Let $\varphi \in L^2(\mathbb{R}^2, \mathbb{H})$ be a quaternion window function. Then we have*

$$\mathcal{G}_{\widetilde{\varphi}}(\widetilde{f})(\omega, b) = \mathcal{G}_\varphi\{f\}(-\omega, -b) \qquad (1.13)$$

where $\widetilde{\varphi}(x) = \varphi(-x)$; $\forall \varphi \in L^2(\mathbb{R}^2, \mathbb{H})$

Proof A direct calculation allows us to obtain for every $f \in L^2(\mathbb{R}^2, \mathbb{H})$

$$\mathcal{G}_{\widetilde{\varphi}}(\widetilde{f})(\omega, b) = \int_{\mathbb{R}^2} e^{-2i\pi x_1\omega_1} f(-x)\overline{\varphi(-(x - b))}e^{-2j\pi x_2\omega_2}dx$$

$$= \int_{\mathbb{R}^2} e^{-2i\pi(-x_1)(-\omega_1)} f(-x)\overline{\varphi(-x - (-b))}e^{-2j\pi(-x_2)(-\omega_2)}dx$$

$$= \mathcal{G}_\varphi\{f\}(-\omega, -b)$$

$$\square$$

For establishing an inversion formula and Plancherel identity for GQFT, we use the fact that the GQFT can be expressed in terms of two-sided quaternion Fourier transform.

$$\mathcal{G}_\varphi\{f\}(\omega, b) = \mathcal{F}_q\{f(.)\varphi(. - b)\}(\omega) \qquad (1.14)$$

Theorem 1.5 (Inversion Formula) *Let φ be a quaternion window function. Then for every function $f \in L^2(\mathbb{R}^2, \mathbb{H})$ can be reconstructed by*

$$f(x) = \frac{1}{\|\varphi\|_2^2} \int_{\mathbb{R}^2} \int_{\mathbb{R}^2} e^{2i\pi x_1 \omega_1} G_\varphi f(w, b) e^{2j\pi x_2 \omega_2} \varphi(x - b) d\omega db \qquad (1.15)$$

Proof We have

$$G_\varphi\{f\}(\omega, b) = \int_{\mathbb{R}^2} e^{-2i\pi x_1 \omega_1} f(x) \overline{\varphi(x - b)} e^{-2j\pi x_2 \omega_2} dx$$

then

$$G_\varphi\{f\}(\omega, b) = \mathcal{F}_q(f(x)\overline{\varphi(x - b)}) \qquad (1.16)$$

Taking the inverse of two-sided QFT of both sides of (1.16) we obtain

$$f(x)\overline{\varphi(x - b)} = \mathcal{F}_q^{-1} G_\varphi f(\omega, b)(x)$$

$$= \int_{\mathbb{R}^2} e^{2i\pi x_1 \omega_1} G_\varphi\{f\}(\omega, b) e^{2j\pi x_2 \omega_2} d\omega \qquad (1.17)$$

Multiplying both sides of (1.17) from the right and integrating with respect to db we get

$$f(x) \int_{\mathbb{R}^2} |\varphi(x - b)|^2 db = \int_{\mathbb{R}^2} \int_{\mathbb{R}^2} e^{2i\pi x_1 \omega_1} G_\varphi f(w, b) e^{2j\pi x_2 \omega_2} \varphi(x - b) d\omega db \qquad (1.18)$$

then,

$$f(x) = \frac{1}{\|\varphi\|_2^2} \int_{\mathbb{R}^2} \int_{\mathbb{R}^2} e^{2i\pi x_1 \omega_1} G_\varphi f(w, b) e^{2j\pi x_2 \omega_2} \varphi(x - b) d\omega db$$

Set $C_\varphi = \|\varphi\|_{\mathbb{R}^2}^2$ and assume that $0 < C_\varphi < \infty$. Then the inversion formula can also written as

$$f(x) = \frac{1}{C_\varphi} \int_{\mathbb{R}^2} \int_{\mathbb{R}^2} e^{2i\pi x_1 \omega_1} G_\varphi f(w, b) e^{2j\pi x_2 \omega_2} \varphi(x - b) d\omega db$$

Theorem 1.6 (Plancherel Theorem) *Let φ be quaternion window function and $f \in L^2(\mathbb{R}^2, \mathbb{H})$, then we have*

$$\|G_\varphi\{f\}\|_2^2 = \|f\|_2^2 \|\varphi\|_2^2 \qquad (1.19)$$

Proof We have

$$\|\mathcal{G}_\varphi\{f\}\|_2^2 = \|\mathcal{F}_q(f(x)\overline{\varphi(x-b)})\|_2^2$$

$$= \|f(x)\overline{\varphi(x-b)}\|_2^2 \tag{1.20}$$

$$= \int_{\mathbb{R}^2} \int_{\mathbb{R}^2} |f(x)|^2 |\varphi(x-b)|^2 dx db$$

$$= \int_{\mathbb{R}^2} |f(x)|^2 dx \int_{\mathbb{R}^2} |\varphi(t)|^2 dt \tag{1.21}$$

$$= \|f\|_2^2 \|\varphi\|_2^2$$

where in line (1.20) we apply the Plancherel theorem of QFT (Theorem 1.2); in (1.21) we use a substitution and Fubini's theorem. $\qquad\square$

1.5 Uncertainty Principles For the GQFT

In this section we demonstrate some versions of uncertainty principles and inequalities for the two-sided quaternion windowed Fourier transform.

1.5.1 Heisenberg Uncertainty Principle

Before proving the Heisenberg uncertainty principle for GQFT, first, we are giving a version of Heisenberg uncertainty for the QFT, which we will use it to demonstrate our result.

Theorem 1.7 Let $f \in L^2(\mathbb{R}^2, \mathbb{H})$ be a quaternion-valued signal such that $x_k f, \frac{\partial}{\partial x_k} f \in L^2(\mathbb{R}^2, \mathbb{H})$ for $k = 1, 2$, then,

$$\left(\int_{\mathbb{R}^2} x_k^2 |f(x)|^2 dx\right)^{\frac{1}{2}} \left(\int_{\mathbb{R}^2} \omega_k^2 |\mathcal{F}_q(f)(\omega)|^2 d\omega\right)^{\frac{1}{2}} \geq \frac{1}{4\pi} \|f\|_2^2 \tag{1.22}$$

To prove this theorem, we need the following result:

Lemma 1.1 Let $f \in L^1 \cap L^2(\mathbb{R}^2, \mathbb{H})$. If $\frac{\partial}{\partial x_k} f$ exist and belong to $L^2(\mathbb{R}^2, \mathbb{H})$ for $k = 1, 2$. Then

$$(2\pi)^2 \int_{\mathbb{R}^2} \omega_k^2 |\mathcal{F}(f(x))(\omega)|^2 d\omega = \int_{\mathbb{R}^2} |\frac{\partial}{\partial x_k} f(x)|^2 dx \tag{1.23}$$

Proof See [6]. $\qquad\square$

We are going to prove the first theorem 1.7.

Proof For $k \in 1, 2$. First, by applying Lemma 1.1 and Plancherel theorem (1.19), we obtain

$$\frac{\int_{\mathbb{R}^2} x_k^2 |f(x)|^2 dx \int_{\mathbb{R}^2} \omega_k^2 |\mathcal{F}_q(f)(\omega)|^2 d^2\omega}{\int_{\mathbb{R}^2} |f(x)|^2 dx \int_{\mathbb{R}^2} |\mathcal{F}_q(f)(\omega)|^2 d\omega}$$

$$= \frac{\frac{1}{(2\pi)^2} \int_{\mathbb{R}^2} x_k^2 |f(x)|^2 dx \int_{\mathbb{R}^2} |\frac{\partial}{\partial x_k} f(x)|^2 d\omega}{\int_{\mathbb{R}^2} |f(x)|^2 dx \int_{\mathbb{R}^2} |\mathcal{F}_q(f)(\omega)|^2 d\omega}$$

$$= \frac{\frac{1}{(2\pi)^2} \int_{\mathbb{R}^2} x_k^2 |f(x)|^2 dx \int_{\mathbb{R}^2} |\frac{\partial}{\partial x_k} f(x)|^2 d\omega}{(\int_{\mathbb{R}^2} |f(x)|^2 dx)^2}$$

$$\geq \frac{1}{16\pi^2} \frac{(\int_{\mathbb{R}^2} (\frac{\partial}{\partial x_k} f(x) x_k \overline{f(x)} + x_k f(x) \frac{\partial}{\partial x_k} \overline{f(x)}) dx)^2}{\|f(x)\|_2^4}$$

$$= \frac{1}{16\pi^2} \frac{(\int_{\mathbb{R}^2} x_k \frac{\partial}{\partial x_k} (f(x) \overline{f(x)}) dx)^2}{\|f(x)\|_2^4}$$

Second, using integration par parts, we further get,

$$= \frac{1}{16\pi^2} \frac{([\int_{\mathbb{R}} x_k |f(x)|^2 dx_l]_{x_k=-\infty}^{x_k=+\infty} - \int_{\mathbb{R}^2} \|f(x)\|^2 dx)^2}{\|f(x)\|_2^4}$$

$$= \frac{1}{16\pi^2}$$

then,

$$\int_{\mathbb{R}^2} x_k^2 |f(x)|^2 dx \int_{\mathbb{R}^2} \omega_k^2 |\mathcal{F}_q(f)(\omega)|^2 d^2\omega \geq \frac{1}{16\pi^2} \int_{\mathbb{R}^2} |f(x)|^2 dx \int_{\mathbb{R}^2} |\mathcal{F}_q(f)(\omega)|^2 d\omega$$
(1.24)

Applying the Plancherel formula (1.2), we obtain our result,

$$\left(\int_{\mathbb{R}^2} x_k^2 |f(x)|^2 dx \right)^{\frac{1}{2}} \left(\int_{\mathbb{R}^2} \omega_k^2 |\mathcal{F}_q(f)(\omega)|^2 d\omega \right)^{\frac{1}{2}} \geq \frac{1}{4\pi} \|f\|_2^2$$

Applying the Plancherel theorem for the QFT (1.2) to the right-hand side of Theorem 1.7, we get the following corollary:

Corollary 1.1 *Under the above assumptions, we have*

$$\left(\int_{\mathbb{R}^2} x_k^2 |\mathcal{F}_q^{-1}\{\mathcal{F}_q(f)\}(x)|^2 dx \right)^{\frac{1}{2}} \left(\int_{\mathbb{R}^2} \omega_k^2 |\mathcal{F}_q(f)(\omega)|^2 d\omega \right)^{\frac{1}{2}} \geq \frac{1}{4\pi} \|\mathcal{F}_q(f)\|_2^2$$
(1.25)

Now, we are going to establish a generalization of the Heisenberg type uncertainty principle for the GQFT.

Theorem 1.8 (Heisenberg for GQFT) *Let* $\varphi \in L^2(\mathbb{R}^2, \mathbb{H})$ *be a quaternion window function and let* $\mathcal{G}_\varphi\{f\} \in L^2(\mathbb{R}^2, \mathbb{H})$ *be the GQFT of* f *such that* $\omega_k \mathcal{G}_\varphi\{f\} \in L^2(\mathbb{R}^2, \mathbb{H}), \quad k = 1, 2.$ *Then for every* $f \in L^2(\mathbb{R}^2, \mathbb{H})$ *we have the following inequality:*

$$\left(\int_{\mathbb{R}^2} x_k^2 |f(x)|^2 dx\right)^{\frac{1}{2}} \left(\int_{\mathbb{R}^2}\int_{\mathbb{R}^2} \omega_k^2 |\mathcal{G}_\varphi\{f\}(\omega, b)|^2 d\omega db\right)^{\frac{1}{2}} \geq \frac{1}{4\pi} \|f\|_2^2 \|\varphi\|_2$$

$$(1.26)$$

In order to prove this theorem, we need to introduce the following lemmas. The first lemma called the Cauchy–Schwartz inequality,

Lemma 1.2 *Let* $f, g \in L^2(\mathbb{R}^2, \mathbb{H})$ *be two quaternion on valued functions. Then the Cauchy–Schwartz inequality takes the form*

$$\left| \int_{\mathbb{R}^2} \overline{f(x)} g(x) dx \right|^2 \leq \int_{\mathbb{R}^2} |f(x)|^2 dx \int_{\mathbb{R}^2} |g(x)|^2 dx$$

Lemma 1.3 *Under the assumptions of Theorem 1.8, we have*

$$\|\varphi\|_2^2 \int_{\mathbb{R}^2} x_k^2 |f(x)|^2 dx = \int_{\mathbb{R}^2}\int_{\mathbb{R}^2} x_k^2 |\mathcal{F}_q^{-1}\{\mathcal{G}_\varphi\{f\}(\omega, b)\}(x)|^2 dx db \qquad (1.27)$$

for $k = 1, 2.$

Proof Applying elementary properties of quaternion, we get

$$\|\varphi\|_2^2 \int_{\mathbb{R}^2} x_k^2 |f(x)|^2 dx = \int_{\mathbb{R}^2} x_k^2 |f(x)|^2 dx \int_{\mathbb{R}^2} |\varphi(x - b)|^2 db$$

$$= \int_{\mathbb{R}^2}\int_{\mathbb{R}^2} x_k^2 |f(x)|^2 |\varphi(x - b)|^2 dx db$$

$$= \int_{\mathbb{R}^2}\int_{\mathbb{R}^2} x_k^2 |f(x)\overline{\varphi(x - b)}|^2 dx db$$

$$= \int_{\mathbb{R}^2}\int_{\mathbb{R}^2} x_k^2 |\mathcal{F}^{-1}(\mathcal{G}_\varphi\{f\}(\omega, b))(x)|^2 dx db$$

$\square$

Now, we are going to prove the Theorem 1.26.

Proof (Of Theorem 1.26) Replacing the QFT of f by the GQFT of the left-hand side of (1.25) in Corollary 1.1, we obtain

$$\left(\int_{\mathbb{R}^2} x_k^2 |\mathcal{F}_q^{-1}\{\mathcal{G}_\varphi\{f\}(\omega, b)\}(x)|^2 dx \right) \left(\int_{\mathbb{R}^2} \omega_k^2 |\mathcal{G}_\varphi\{f\}(\omega, b)|^2 d\omega \right) \geq$$

$$\frac{1}{16\pi^2} \left(\int_{\mathbb{R}^2} |\mathcal{G}_\varphi f(\omega, b)|^2 d\omega \right)^2 \quad (1.28)$$

we have

$$\mathcal{F}^{-1}(\mathcal{G}_\varphi\{f\}(\omega, b))(x) = f(x)\overline{\varphi(x - b)}$$

Taking the square root on both sides of (1.28) and integrating both sides with respect to db we get

$$\int_{\mathbb{R}^2} \left(\int_{\mathbb{R}^2} x_k^2 |\mathcal{F}_q^{-1}\{\mathcal{G}_\varphi\{f\}(\omega, b)\}(x)|^2 dx \right)^{\frac{1}{2}} \left(\int_{\mathbb{R}^2} \omega_k^2 |\mathcal{G}_\varphi\{f\}(\omega, b)|^2 d\omega \right)^{\frac{1}{2}} db \geq$$

$$\frac{1}{4\pi} \int_{\mathbb{R}^2} \int_{\mathbb{R}^2} |\mathcal{G}_\varphi\{f\}(\omega, b)|^2 d\omega db$$

$$(1.29)$$

Applying the Cauchy–Schwartz inequality (Lemma 1.2) to the left-hand side of (1.29) we obtain

$$\left(\int_{\mathbb{R}^2} \int_{\mathbb{R}^2} x_k^2 |\mathcal{F}_q^{-1}\{\mathcal{G}_\varphi\{f\}(\omega, b)\}(x)|^2 dx db \right)^{\frac{1}{2}} \left(\int_{\mathbb{R}^2} \int_{\mathbb{R}^2} \omega_k^2 \mathcal{G}_\varphi\{f\}(\omega, b)|^2 d\omega db \right)^{\frac{1}{2}}$$

$$\geq \frac{1}{4\pi} \int_{\mathbb{R}^2} \int_{\mathbb{R}^2} |\mathcal{G}_\varphi\{f\}(\omega, b)|^2 d\omega db$$

$$(1.30)$$

Using Lemma 1.27 into the second term on the left-hand side of (1.30), and using the Plancherel formula (1.6) into the right-hand side of (1.30), we obtain that

$$\left(\|\varphi\|_2^2 \int_{\mathbb{R}^2} x_k^2 |f(x)|^2 dx \right)^{\frac{1}{2}} \left(\int_{\mathbb{R}^2} \int_{\mathbb{R}^2} \omega_k^2 |\mathcal{G}_\varphi\{f\}(\omega, b)|^2 d\omega db \right)^{\frac{1}{2}} \geq \frac{1}{4\pi} \|f\|_2^2 \|\varphi\|_2^2$$

$$(1.31)$$

Now, simplifying both sides of (1.31) by $\|\varphi\|_2$, we get our result.

1.5.2 Logarithmic Inequality

Definition 1.4 A couple $\alpha = (\alpha_1, \alpha_2)$ of non-negative integers is called a multi-index. One denotes

$$|\alpha| = \alpha_1 + \alpha_2 \ \text{ and } \ \alpha! = \alpha_1!\alpha_2!$$

and, for $x \in \mathbb{R}^2$

$$x^\alpha = x_1^{\alpha_1} x_2^{\alpha_2}$$

Derivatives are conveniently expressed by multi-indices

$$\partial^\alpha = \frac{\partial^{|\alpha|}}{\partial x_1^{\alpha_1} \partial x_2^{\alpha_2}}$$

Next, we obtain the Schwartz space as [10]

$$S(\mathbb{R}^2, \mathbb{H}) = \{f \in C^\infty(\mathbb{R}^2, \mathbb{H}) : sup_{x \in \mathbb{R}^2}(1 + |x|^k)|\partial^\alpha f(x)| < \infty\}$$

where $C^\infty(\mathbb{R}^2, \mathbb{H})$ is the set of smooth function from $\mathbb{R}^2$ to $\mathbb{H}$.

We have the logarithmic uncertainty principle for the QFT [11] as follows:

Theorem 1.9 (QFT Logarithmic Uncertainty Principle) *For $f \in S(\mathbb{R}^2, \mathbb{H})$, we have*

$$\int_{\mathbb{R}^2} ln|x||f(x)|^2 dx + \int_{\mathbb{R}^2} ln|\omega||\mathcal{F}_q\{f\}(\omega)|^2 d\omega \geq \left(\frac{\Gamma'(t)}{\Gamma(t)} - ln\pi\right) \int_{\mathbb{R}^2} |f(x)|^2 dx$$

$$(1.32)$$

where $\Gamma'(t) = \left(\frac{d}{dt}\right)$ and $\Gamma(t)$ is Gamma function.

Remark 1.1 If we apply Plancherel theorem for QFT (1.2) to the right-hand side of (1.32), we get

$$\int_{\mathbb{R}^2} ln|x||f(x)|^2 dx + \int_{\mathbb{R}^2} ln|\omega||\mathcal{F}_q\{f\}(\omega)|^2 d\omega \geq \left(\frac{\Gamma'(t)}{\Gamma(t)} - ln\pi\right) \int_{\mathbb{R}^2} |\mathcal{F}_q\{f\}(\omega)|^2 d\omega dx$$

$$(1.33)$$

Lemma 1.4 *Let $\varphi \in S(\mathbb{R}^2, \mathbb{H})$ a windowed quaternionic function and $f \in S(\mathbb{R}^2, \mathbb{H})$. We have*

$$\int_{\mathbb{R}^2} \int_{\mathbb{R}^2} ln|x||\mathcal{F}_q^{-1}\{G_\varphi f(\omega, b)\}(x)|^2 dx db = \|\varphi\|^2_{L^2(\mathbb{R}^2, \mathbb{H})} \int_{\mathbb{R}^2} ln|x||f(x)|^2 dx$$

$$(1.34)$$

Proof By a simple calculation we get

$$\int_{\mathbb{R}^2} \int_{\mathbb{R}^2} ln|x||\mathcal{F}_q^{-1}\{G_\varphi f(\omega, b)\}(x)|^2 dx db = \int_{\mathbb{R}^2} \int_{\mathbb{R}^2} ln|x||f(x)\overline{\varphi(x - b)}|^2 dx db$$

$$= \int_{\mathbb{R}^2} \int_{\mathbb{R}^2} ln|x||f(x)|^2 |\varphi(x - b)|^2 dx db$$

$$= \int_{\mathbb{R}^2} ln|x||f(x)|^2 \int_{\mathbb{R}^2} |\varphi(x-b)|^2 db dx$$

$$= \|\varphi\|^2_{L^2(\mathbb{R}^2, \mathbb{H})} \int_{\mathbb{R}^2} ln|x||f(x)|^2 dx$$

Corollary 1.2 *For* $f \in \mathcal{S}(\mathbb{R}^2, \mathbb{H})$, *and* $\varphi \in \mathcal{S}(\mathbb{R}^2, \mathbb{H})$, *we have*

$$\int_{\mathbb{R}^2} ln|x||\mathcal{F}_q^{-1}\mathcal{F}_q(f)(x)|^2 dx + \int_{\mathbb{R}^2} ln|\omega||\mathcal{F}_q\{f\}(\omega)|^2 d\omega \geq$$

$$\left(\frac{\Gamma'(t)}{\Gamma(t)} - ln\pi\right) \int_{\mathbb{R}^2} |\mathcal{F}_q(\omega)|^2 d\omega \qquad (1.35)$$

Theorem 1.10 *Let* $f \in \mathcal{S}(\mathbb{R}^2, \mathbb{H})$ *and* $\varphi \in \mathcal{S}(\mathbb{R}^2, \mathbb{H})$ *a quaternion windowed function, we have the following algorithmic inequality:*

$$\|\varphi\|^2_{L^2(\mathbb{R}^2, \mathbb{H})} \int_{\mathbb{R}^2} ln|x||f(x)|^2 dx + \int_{\mathbb{R}^2} \int_{\mathbb{R}^2} ln|\omega||\mathcal{G}_\varphi f(\omega, b)|^2 d\omega db \geq$$

$$\|\varphi\|^2_{L^2(\mathbb{R}^2, \mathbb{H})} \left(\frac{\Gamma'(t)}{\Gamma(t)} - ln\pi\right) \int_{\mathbb{R}^2} \int_{\mathbb{R}^2} |f(x)|^2 dx \qquad (1.36)$$

Proof For classical two-sided quaternionic Fourier transform, by Theorem 1.9,

$$\int_{\mathbb{R}^2} ln|x||f(x)|^2 dx + \int_{\mathbb{R}^2} ln|\omega||\mathcal{F}_q\{f\}(\omega)|^2 d\omega \geq \left(\frac{\Gamma'(t)}{\Gamma(t)} - ln\pi\right) \int_{\mathbb{R}^2} |f(x)|^2 dx, \qquad (1.37)$$

we replace f by $\mathcal{G}_\varphi f$ on both sides of (1.37), we get

$$\int_{\mathbb{R}^2} ln|\omega||\mathcal{G}_\varphi f(\omega, b)|^2 d\omega + \int_{\mathbb{R}^2} ln|x||\mathcal{F}_q\{\mathcal{G}_\varphi f\}(x)|^2 dx \geq$$

$$\left(\frac{\Gamma'(t)}{\Gamma(t)} - ln\pi\right) \int_{\mathbb{R}^2} |\mathcal{G}_\varphi f(\omega, b)|^2 dx \qquad (1.38)$$

Integrating both sides of this equation with respect to db, we obtain

$$\int_{\mathbb{R}^2} \int_{\mathbb{R}^2} ln|\omega||\mathcal{G}_\varphi f(\omega, b)|^2 d\omega db + \int_{\mathbb{R}^2} \int_{\mathbb{R}^2} ln|x||\mathcal{F}_q\{\mathcal{G}_\varphi f\}(x)|^2 dx db \geq$$

$$\left(\frac{\Gamma'(t)}{\Gamma(t)} - ln\pi\right) \int_{\mathbb{R}^2} \int_{\mathbb{R}^2} |\mathcal{G}_\varphi f(\omega, b)|^2 dx db \qquad (1.39)$$

Applying Lemma 1.4 into the second term on the left-hand side of (1.39), yields

$$\int_{\mathbb{R}^2} \int_{\mathbb{R}^2} ln|\omega| |\mathcal{G}_\varphi f(\omega, b)|^2 d\omega db + \|\varphi\|^2_{L^2(\mathbb{R}^2, \mathbb{H})} \int_{\mathbb{R}^2} ln|x| |f(x)|^2 dx \qquad (1.40)$$

$$\geq \left(\frac{\Gamma'(t)}{\Gamma(t)} - ln\pi \right) \|\varphi\|^2_{L^2(\mathbb{R}^2, \mathbb{H})} \int_{\mathbb{R}^2} \int_{\mathbb{R}^2} |\mathcal{G}_\varphi f(\omega, b)|^2 dx db \qquad (1.41)$$

On applying the Plancherel formula in the right side of (1.40), we obtain our desired result,

$$\|\varphi\|^2_{L^2(\mathbb{R}^2, \mathbb{H})} \int_{\mathbb{R}^2} ln|x| |f(x)|^2 dx + \int_{\mathbb{R}^2} \int_{\mathbb{R}^2} ln|\omega| |\mathcal{G}_\varphi f(\omega, b)|^2 d\omega db$$

$$\geq \|\varphi\|^2_{L^2(\mathbb{R}^2, \mathbb{H})} \left(\frac{\Gamma'(t)}{\Gamma(t)} - ln\pi \right) \int_{\mathbb{R}^2} \int_{\mathbb{R}^2} |f(x)|^2 dx$$

References

1. Assefa, D., Mansinha, L., Tiampo, K.F. et al.: (2010). Local quaternion Fourier transform and color image texture analysis. *Signal Process*, 90(6), 1825–1835.
2. Bahri, M., Hitzer, E., Ashino, R., Vaillancourt, R. (2010). Windowed Fourier transform of two-dimensional quaternionic signals. *Applied Mathematics and Computation*, 216, 2366–2379.
3. Bayro-Corrochano, E., Trujillo, N., Naranjo, M. (2007). Quaternion Fourier descriptors for the preprocessing and recognition of spoken words using images of spatiotemporal representations. *J. Math. Imaging Vision*. 28(2):179–190.
4. Bülow, T. (1999). Hypercomplex spectral signal representations for the processing and analysis of images. *Kiel: Universität Kiel. Institut für Informatik und Praktische Mathematik* Writing, 44(3), 213–245.
5. Bülow, T., Felsberg, M., Sommer, G. (2001). Non-commutative hypercomplex Fourier transforms of multidimensional signals. *G. Sommer (Ed). Geom. Comp. with Cliff. Alg., theor. Found. and Appl. in comp.* Vision and Robotics, ed. Springer, 187–207.
6. El Haoui, Y., Fahlaoui, S. (2017). The uncertainty principle for the two-sided quaternion Fourier transform. *S. Mediterr. J. Math.* Writing, 14(6): 221.
7. Ell, TA., Sangwine, S.J. (2007). Hypercomplex Fourier transforms of color images. *IEEE Trans Image Process*, 16(1), 22–35.
8. Ell, T.A. (1993).Quaternion-Fourier transforms for analysis of two-dimensional linear time-invariant partial differential systems. *Proceeding of the 32nd Conference on Decision and Control* San Antonio, Texas, 1830–1841.
9. Fu, Y., Kähler, U., Cerejeiras, P. (2012). The Balian-Low Theorem for the Windowed Quaternionic Fourier Transform. *Adv. Appl. Clifford Algebras*, 22, 1025.
10. Kit Ian, K. and Morais, J. (2014). Asymptotic behaviour of the quaternion linear canonical transform and the Bochner-Minlos theorem. *Applied Mathematics and Computation*, 247(15), 675–688.

11. Li-Ping, C., Kit Ian, K., Ming-Sheng, L. (2015). Pitt's inequality and the uncertainty principle associated with the quaternion Fourier transform. *J. Math. Ana. App* Writing, 423, 681–700.
12. Pei, S.C., Ding, J.J., Chang, J.H.r. (2001). Efficient implementation of quaternion Fourier transform, convolution, and correlation by 2-D complex FFT. *JIEEE Trans.* Signal Process, 49(11), 2783–2797.

Chapter 2
Reflected Backward SDEs in a Convex Polyhedron

Khadija Akdim

Abstract A backward stochastic differential equation is forced to stay within a d-dimensional bounded convex polyhedral domain, thanks to the action of oblique reflecting process at the boundary. The Lipschitz continuity on the reflection directions together with the Lipschitz continuity of the drift gives the existence and uniqueness of the solution.

Keywords Vector field · Gaussian process · Random field · Covariance operator

2.1 Introduction

This paper is concerned with the oblique reflection problems that have application in queuing theory. We consider a backward stochastic differential equation in the case of d-dimensional convex bounded region with oblique reflecting conditions. Using a modification of the framework of Ouknine [8], Menaldi and Robin [7] as well Gegout-Petit and Pardoux [3], we prove the existence and uniqueness of the solution by a contraction mapping argument, our result generalize to oblique reflection case Theorem 2.1 given in [8]. Let us mention that the penalization approach for reflected backward stochastic differential equations without jumps is given in Gegout-Petit and Pardoux [3], with jumps and normal reflection in Ouknine [8]. This kind of problem, that has applications in queuing and storage theory, has also been studied in the case of forward stochastic differential equations by a direct approach based on the Skorokhod problem in Watanabé [11], Tanaka [10], Chaleyat-Maurel et al. [1], Menaldi and Robin [7] for reflected forward stochastic differential equations driven by Lévy processes; their results hold under the restriction that the jumps of the Lévy process place the solution process inside the domain, Lions and Sznitmann [6] and

K. Akdim (✉)
Department of Mathematics, Faculty of Sciences and Techniques, Cadi Ayyad University, Marrakech, Morocco
e-mail: k.akdim@uca.ac.ma

© Springer Nature Switzerland AG 2020
S. Dos Santos et al. (eds.), *Recent Advances in Mathematics and Technology*, Applied and Numerical Harmonic Analysis, https://doi.org/10.1007/978-3-030-35202-8_2

also in some books like Liptser and Shiryayev [5] with some application in queueing models, Ikeda and Watanabe [4]. We treat only the case of regular convex domain; the case of general domain is an open and interesting problem.

The aim of this article is to study a reflected backward stochastic differential equation (RBSDE) in convex polyhedral domains with oblique reflection at the boundary. The drift vector and the reflection matrix can be time and space dependent; existence and uniqueness are established under Lipschitz continuity on the reflection matrix and Lipschitz continuity of the drift.

The rest of the paper is organized as follows. Next section is devoted to the notations we use during this text, setting of the problem, some known proprieties of regular convex domain. In the last section, we claim our main that we prove in the sequel.

2.2 Statement of the Problem

2.2.1 Notations

Let $(W_t, t \in [0, T])$ be a d-dimensional standard Brownian motion process defined on a probability space $(\Omega, \mathcal{F}, \mathbb{P})$; let $(\mathcal{F}_t, t \in [0, T])$ be the natural filtration generated by $(W_t, t \in [0, T])$ with $\mathcal{F}_0$ containing all $\mathbb{P}$-null sets.

Let us introduce some notation:

1. A subset $\mathcal{O}$ of $\mathbb{R}^d$ such that $\mathcal{O}$ is an bounded convex polyhedral domain assumed to be non-empty, $\mathcal{O} = \{x \in \mathbb{R}^d : n_i.x > c_i, i = 1, \ldots, N\}$; where n_i is a unit vector normal to the hyperplane H_i and $n_i.x$ denotes the inner product of the vector n_i and x.
2. The hyperplane $H_i = \{x \in \mathbb{R}^d : n_i.x = c_i\}$, $i = 1, \ldots, N$.
3. The faces F_i of the polyhedron $\overline{\mathcal{O}}$ which has dimension $n - 1$, with

$$F_i = \{x \in \overline{\mathcal{O}} : n_i.x = c_i\}$$

$i = 1, \ldots, N$. We define:
4. $\mathcal{M}_d^2$ the set of $\mathcal{F}_t$-progressively measurable process (w_t) taking values in $\mathbb{R}^{d \times d}$ such that $\mathbb{E}\left[\int_0^T |w_s|^2\, ds\right] < \infty$.
5. $\mathcal{S}^2$ be the set of $\mathcal{F}_t$-progressively measurable continuous process (v_t) taking values in $\mathbb{R}^d$ such that $\mathbb{E}[\sup_{0 \le t \le T} |v_t|^2] < \infty$.

Finally, we define

$$\mathbf{B}_d^{(2,N)} \triangleq \mathcal{S}^2 \times \mathcal{M}_d^2 \times \mathbb{R}^N. \tag{2.1}$$

Now, we are given the following objects:

(H_1) a terminal value $\xi \in \mathcal{L}^2(\Omega, \mathcal{F}_T, \mathbb{P})$ such that $\xi \in \overline{\mathcal{O}}$;
(H_2) a function process f, which is map

$$f : \Omega \times [0, T] \times \mathbb{R}^d \longrightarrow \mathbb{R}^d \tag{2.2}$$

such that

(a) $\forall y \in \mathbb{R}^d$: $(\omega, t) \longrightarrow f(\omega, t, y)$ is $\mathcal{F}_t$ − progressively measurable.

(b) $\mathbb{E} \int_0^T \| f(\omega, t, 0) \|^2 \, dt < +\infty,$

(c) $\exists \, \alpha > 0$, such that for any y and $y' \in \mathbb{R}^d$ and $(\omega, t) \in \Omega \times [0, T]$

$$\| f(\omega, t, y) - f(\omega, t, y') \| \le \alpha \, \| y - y' \| \, ;$$

(H_3)   the vector v_k, $k = 1, \dots, N$, is defined for each $x \in H_k$. By this vector is meant the reflection direction at the point x, $x \in H_k$. We shall need to extend vector-valued function v_k for arbitrary values $x \in \mathbb{R}^d$. Trivially, we define $v_k(t, x) = v_k\left(t, \, pr_{H_k}(x)\right)$, $x \in \mathbb{R}^d$ which is map

$$v_k : \Omega \times [0, T] \times \overline{\mathcal{O}} \longrightarrow \mathbb{R}^d \tag{2.3}$$

such that

(a) $n_k . v_k(t, x) = 1$, $x \in H_k$, $k = 1, \dots, N$, this condition of normalization of the length of the vector v_k will considerably simplify the presentation of the main results;

(b) $\exists \, \gamma > 0$, for all $1 \le k \le N$ such that for any $x, \, x' \in \overline{\mathcal{O}}$; $(\omega, t) \in \Omega \times [0, T]$;

$$\| v_k(w, t, x) - v_k(w, t, x') \| \le \gamma \, \| x - x' \| , \tag{2.4}$$

(H_4)   for each $x \in \partial\mathcal{O}$ we require the existence of positive numbers $a_1, \dots a_N, \lambda$ with $0 < \lambda < 1$, see, for example, [9], such that we have

$$\sum_{i=1, i \neq k}^{N} a_i |n_i . v_k(\omega, t, x)| < \lambda a_k, \ k = 1, \dots, N. \tag{2.5}$$

2.2.2 Main Definition

Let us now introduce our BSDE in a convex polyhedron with oblique reflection.

Definition 2.1 The solution of our BSDE associated with $(\xi, f, \mathcal{O}, v)$ is a triple $\{(Y_t, Z_t, K_t), \ 0 \le t \le T\}$ of $\mathcal{F}_t$-progressively measurable processes taking values in $\mathbb{R}^d$, $\mathbb{R}^{d \times d}$, and $\mathbb{R}^N$, respectively, and satisfying:

1. $Z \in \mathcal{M}_d^2$ is a predictable processes, in particular

$$\mathbb{E}\left(\int_0^T \|Z_t\|^2 \, dt\right) < +\infty, \tag{2.6}$$

2. $Y \in \mathcal{S}^2$ with instantaneous reflection at the boundary $\partial \mathcal{O}$ in the direction
   $v = (v_i)_{1 \le i \le N}$;
3. for every $0 \le t \le T$

$$Y_t = \xi + \int_t^T f(s, Y_s) ds + \sum_{i=1}^N \int_t^T v_i(s, Y_s) dK_s^i - \int_t^T Z_s dW_s;$$

4. $Y_t \in \overline{\mathcal{O}}$, for all $0 \le t \le T$;
5. $K \in \mathcal{S}^2$ such that every component-function $K_t^i, i = 1, \ldots, N$, is nondecreasing
   continuous process, with $K_0^i = 0, i = 1, \ldots, N$, and

$$K_t^i = \int_0^t \varkappa(n_i . Y_s = c_i) dK_s^i \,, \quad i = 1, \ldots, N, \tag{2.7}$$

where $\varkappa(n_i . Y_s = c_i)$ denotes the characteristic function of the face F_i.

2.3   Existence and Uniqueness

Proposition 2.1 *Assume* $(H_1) - (H_4)$ *and suppose* $(Y_t, Z_t, K_t)_{0 \le t \le T}$ *solve RBSDE*
$(\xi, f, \mathcal{O}, v)$. *Then*

$$0 \le dK_t^i \le \left((\mathbf{I} - \mathbf{V})^{-1} C\right)_i dt \; i = 1, \ldots N. \tag{2.8}$$

Such that $\mathbf{V} = (v_{ij})_{0 \le ij \le N}$ *for* $i, j \in \{1, \ldots, N\}$ *where* v_{ij} *is a constant for* $i \ne k$
and $v_{ii} = 0$, *in the other hand* $C = (c_i)_{0 \le i \le N}$ *for all* $i = 1, \ldots, N$ *where* c_i *is a*
constants.

   In particular, for each $i = 1, \ldots N$, K^i *is absolutely continuous with respect to*
the Lebesgue measure.

Proof By (iv) we have

$$Y_t = \xi + \int_t^T f(s, Y_s) ds + \sum_{k=1}^N \int_t^T v_k(s, Y_s) dK_s^k - \int_t^T Z_s dW_s.$$

Taking the inner product with vectors $n_i, i = 1, \ldots N$, we obtain the following
auxiliary problem:

$$n_i.Y_t - c_i = n_i.\xi - c_i + \int_t^T n_i.f(s, Y_s)ds + \sum_{k=1,k\neq i}^{N} \int_t^T n_i.v_k(s, Y_s)dK_s^k$$

$$+ \int_t^T n_i.v_i(s, Y_s)dK_s^i - \int_t^T n_i.Z_s dW_s.$$

By $(H_3) - (a)$ we get

$$n_i.Y_t - c_i = n_i.\xi - c_i + \int_t^T n_i.f(s, Y_s)ds + \sum_{k=1,k\neq i}^{N} \int_t^T n_i.v_k(s, Y_s)dK_s^k + K_T^i - K_t^i$$

$$- \int_t^T n_i.Z_s dW_s. \tag{2.9}$$

Proceeding as in the proof of Proposition 4.2 and Remark 4.3 of El Karoui and al [2] we get for each $i = 1, \ldots N$:

$$0 \leq dK_t^i \leq 1_{\{n_i.Y_t - c_i\}}(Y_t^i) \left\{ |n_i.f(s, Y_s)|dt + \sum_{k=1,k\neq i}^{N} |n_i.v_k(s, Y_s)|dK_s^k \right\}$$

$$\leq |n_i.f(s, Y_s)|dt + \sum_{k=1,k\neq i}^{N} |n_i.v_k(s, Y_s)|dK_s^k.$$

Since the set $\overline{\mathcal{O}}$ is a compact subset of $\mathbb{R}^d$, it is not a serious restriction to suppose that the functions $n_i.f(s, Y_s)$ and $n_i.v_k(s, Y_s)$, $k = 1, \ldots, N$ are bounded in the set $\overline{\mathcal{O}}$, respectively, by a constant c_i and v_{ik}.
Hence we can write

$$0 \leq dK_t^i \leq c_i dt + v_{ik} \sum_{k=1,k\neq i}^{N} dK_s^k.$$

So we obtain

$$dK_t^i - v_{ik} \sum_{k=1,k\neq i}^{N} dK_s^k \leq c_i dt, \quad i = 1, \ldots, N.$$

This inequality can be expressed as

$$\{(\mathbf{I} - \mathbf{V})dK\}_i(t) \leq (C)_i dt, \quad i = 1, \ldots, N.$$

$\square$

If we assume that $\sigma(V) < 1$, where $\sigma(V)$ denotes the spectral radius of V, therefore,

$$(I - V)^{-1} = I + V + V^2 + V^3 + \ldots$$

Which complete the proof.

Clearly for $i = 1, \ldots, N$ K^i is of bounded variations a.s.; in fact, by the preceding Proposition (2.1) K^i are absolutely continuous, let $D_t = (D_t^1, \ldots D_t^N)_{0 \leq t \leq T}$ denote the matrix valued $\mathcal{F}_t$-progressively measurable process of their derivatives where

$$0 \leq D_t^i \leq \left\{(\mathbf{I} - \mathbf{V})^{-1} C\right\}_i \quad \text{a.s and} \quad K_t^i = \int_0^t D_s^i ds \quad \text{for all } 0 \leq t \leq T.$$

Now, let $\mathcal{H}$ stands for the space of all $\mathcal{F}_t$-progressively measurable, continuous pairs of processes $(Y_t)_{0 \leq t \leq T}$ and $(K_t)_{0 \leq t \leq T}$ such that

1. $n_i.Y_t - c_i \geq 0$ for all $0 \leq t \leq T$,
2. $K_0^i = 0$, K^i is nondecreasing and can increase only when $n_i.Y_t = c_i$,
3. $\mathbb{E}\left(\sum_{i=1}^N \int_0^T e^{\beta t} a_i |n_i.Y_t| dt\right) < \infty$,
4. $\mathbb{E}\left(\sum_{i=1}^N \int_0^T e^{\beta t} a_i \varphi_t(K^i) dt\right) < \infty$,

where $\varphi_t(x)$ denotes the total variation of x over $[t, T]$ and $\beta > 0$ is a fixed constant which will be chosen suitably later.
For $(Y, K), (Y', K') \in \mathcal{H}$, we define the metric

$$d\left((Y, K), (Y', K')\right) = \mathbb{E} \sum_{i=1}^N \int_0^T e^{\beta t} a_i |n_i.Y_t - n_i Y_t'| dt$$

$$+ \mathbb{E} \sum_{i=1}^N \int_0^T e^{\beta t} a_i \varphi_t(K^i - K'^i) dt. \qquad (2.10)$$

It is not difficult to see that $(\mathcal{H}, N)$ is a complete metric space.

Before starting our main result, let us remark that if $(Y, K), (Y', K') \in \mathcal{H}$ with D^i, D'^i being, respectively, the derivatives of K^i, K'^i then

$$\varphi_t(K^i - K'^i) = \int_t^T |D_s^i - D_s'^i| ds.$$

If we consider the space $\mathbb{R}^N$ with the norm

$$\|y\| = \sum_{i=1}^N a_i |y^i|,$$

therefore, using integration by parts in (2.10), we have

$$
d\left((Y, K), (Y', K')\right) = \mathbb{E} \sum_{i=1}^{N} \int_0^T e^{\beta t} a_i |n_i.Y_t - n_i.Y_t'| dt
$$

$$
+ \mathbb{E} \sum_{i=1}^{N} \int_0^T \frac{(e^{\beta t} - 1)}{\beta} a_i |D_t^i - D_t'^i| dt
$$

$$
= \mathbb{E} \int_0^T e^{\beta t} \|Y_t - Y_t'\| dt + \mathbb{E} \int_0^T \frac{(e^{\beta t} - 1)}{\beta} \|D_t - D_t'\| dt.
$$

$$(2.11)$$

Remark 2.1 As the norm $\|y\|$ is equivalent to the Euclidean norm in $\mathbb{R}^N$. So, we may assume that the Lipschitz continuity is with respect to this norm.

Now, let us state the principal result of this paper:

Theorem 2.1 *Under the assumptions on data, the RBSDE associated with $(\xi, f, \mathcal{O}, v)$ has one and only one solution $\{(Y_t, Z_t, K_t) \ 0 \le t \le T\}$.*

Proof Let f and v satisfy, respectively, $(H_2) - (c)$ and $(H_3) - (b)$. For a given process $(Y, K) \in \mathcal{H}$, we consider the following reflected BSDE:

$$
n_i.Y_t - c_i = n_i.\xi - c_i + \int_t^T n_i.f(s, Y_s) ds + \sum_{k=1, l \ne k}^{N} \int_t^T n_i.v_k(s, Y_s) D_s^k ds
$$

$$
+ K_T^i - K_t^i - \int_t^T n_i.Z_s dW_s, \quad (2.12)
$$

if we set

$$
g_i(t) = n_i.f(t, Y_t) + \sum_{k=1, l \ne k}^{N} n_i.v_k(t, Y_t) D_s^k \ i = 1, \ldots .N; \quad (2.13)
$$

we get the following one-dimensional RBSDE:

1. for $i = 1, \ldots .N$

$$
n_i.Y_t - c_i = n_i.\xi - c_i + \int_t^T g_i(s) ds + K_T^i - K_t^i - \int_t^T n_i.Z_s dW_s, \quad (2.14)
$$

2. $n_i.Y_t - c_i \ge 0$ for all $0 \le t \le T$,
3. $K_0^i = 0$, K^i is nondecreasing, K^i can increase only when $n_i.Y_t = c_i$.

Then under our assumption and in view of Proposition 5.1 of [2], for all $1 \leq i \leq N$ there exists a unique vector $(n_i.\overline{Y}_t, n_i.\overline{Z}_t, \overline{K}_t^i)$ solves the following system:

$$n_i.\overline{Y}_t - c_i = n_i.\xi - c_i + \int_t^T g_i(s)ds + \overline{K}_T^i - \overline{K}_t^i$$

$$- \int_t^T n_i.\overline{Z}_s dW_s, \quad k = 1,\ldots.N. \tag{2.15}$$

Now, introduce the mapping ϕ from $\mathcal{H}$ into itself as follows. Given (Y, K); $(Y', K') \in \mathcal{H}$, let $(\overline{Y}, \overline{K}) = \phi(Y, K)$ and $(\overline{Y}', \overline{K}') = \phi(Y', K')$, where $(\overline{Y}, \overline{K}), (\overline{Y}', \overline{K}')$ be obtained by solving the associated one-dimensional problems (2.15); see (2.13). So there exist matrix valued $\mathcal{F}_t$-progressively measurable square integrable processes $\overline{Z}, \overline{Z}'$ such that

$$n_i.\overline{Y}_t - c_i = n_i.\xi - c_i + \int_t^T n_i.f(s, Y_s)ds$$

$$+ \sum_{k=1,k\neq i}^N \int_t^T n_i.v_k(s, Y_s)D_s^k ds + \overline{K}_T^i - \overline{K}_t^i$$

$$- \int_t^T n_i.\overline{Z}_s dW_s, \quad k = 1,\ldots.N \tag{2.16}$$

and

$$n_i.\overline{Y}_t' - c_i = n_i.\xi - c_i + \int_t^T n_i.f(s, Y_s')ds$$

$$+ \sum_{k=1,k\neq i}^N \int_t^T n_i.v_k(s, Y_s')D_s'^k ds + \overline{K}_T'^i - \overline{K}_t'^i$$

$$- \int_t^T n_i.\overline{Z}_s' dW_s, \quad k = 1,\ldots.N, \tag{2.17}$$

where $dK_t'^i = D_t'^i dt$ and $d\overline{K}_t'^i = \overline{D}_t'^i dt$
Let $\widehat{Y} = Y - Y'$, $\widehat{K} = K - K'$.
It follows from Itô's formula that for any $\beta > 0$,

$$\mathbb{E}\int_0^T e^{\beta s}\beta|\widehat{n_i.\overline{Y}}_s|ds + \mathbb{E}\int_0^T e^{\beta s}\beta\varphi_s(\widehat{\overline{K}^i})ds$$

$$= \mathbb{E}\int_0^T e^{\beta s}\beta|\widehat{n_i.Y}_s|ds + \mathbb{E}\int_0^T (e^{\beta s} - 1)|\widehat{D_s^i}|ds$$

$$\leq \mathbb{E} \int_0^T (e^{\beta s} - 1) \left| n_i . f(s, Y_s) - n_i . f(s, Y_s') \right| ds$$

$$+ \mathbb{E} \int_0^T (e^{\beta s} - 1) \left| \sum_{k=1, k \neq i}^N n_i . v_k(s, Y_s) D_s^k - \sum_{k=1, k \neq i}^N n_i . v_k(s, Y_t') D_s'^k \right| ds$$

$$\leq \mathbb{E} \int_0^T (e^{\beta s} - 1) \left| n_i . f(s, Y_s,) - n_i . f(s, Y_s') \right| ds$$

$$+ \mathbb{E} \int_0^T (e^{\beta s} - 1) \sum_{k=1, k \neq i}^N \left| n_i . v_k(s, Y_s) - n_i . v_k(s, Y_s') \right| D_s^k ds$$

$$+ \mathbb{E} \int_0^T (e^{\beta s} - 1) \sum_{k=1, k \neq i}^N \left| n_i . v_k(s, Y_s') \right| |D_s^k - D_s'^k| ds.$$

Multiplying the above inequality by a_i and then taking the sum, we deduce that

$$\beta d \left((\overline{Y}, \overline{K}), (\overline{Y}', \overline{K}') \right)$$

$$\leq \mathbb{E} \int_0^T (e^{\beta s} - 1) \| f(s, Y_s,) - f(s, Y_s') \| ds$$

$$+ \mathbb{E} \int_0^T (e^{\beta s} - 1) \sum_{i=1}^N a_i \left(\sum_{k=1, k \neq i}^N \left| n_i . v_k(s, Y_s) - n_i . v_k(s, Y_s') \right| D_s^k \right) ds$$

$$+ \mathbb{E} \int_0^T (e^{\beta s} - 1) \sum_{i=1}^N a_i \left(\sum_{k=1, k \neq i}^N \left| n_i . v_k(s, Y_s') \right| |D_s^k - D_s'^k| \right) ds$$

$$\leq \mathbb{E} \int_0^T (e^{\beta s} - 1) \| f(s, Y_s,) - f(s, Y_s') \| ds$$

$$+ \mathbb{E} \int_0^T (e^{\beta s} - 1) \sum_{k=1}^N \| v_k(s, Y_s) - v_k(s, Y_s') \| D_s^k ds$$

$$+ \mathbb{E} \int_0^T (e^{\beta s} - 1) \sum_{k=1}^N \left(\sum_{i=1, i \neq k}^N a_i \left| n_i . v_k(s, Y_s') \right| \right) |D_s^k - D_s'^k| ds.$$

In view of the Proposition (2.1) we get $\sum_{k=1}^N D^k \leq \sum_{k=1}^N \left((\mathbf{I} - \mathbf{V})^{-1} C \right)_k \leq C$ for some constant C and by using the Lipschitz condition we get

$$\beta d\left((\overline{Y}, \overline{K}), (\overline{Y}', \overline{K}')\right) \leq (\alpha + C\gamma)\mathbb{E}\int_0^T (e^{\beta s} - 1)\|\widehat{Y}_s\|ds$$

$$+ \mathbb{E}\int_0^T (e^{\beta s} - 1)\sum_{k=1}^N \left(\sum_{i=1,i\neq k}^N a_i \left|n_i.v_k(s, Y_s')\right|\right)|D_s^k - D_s'^k|ds,$$

by using the inequalities (2.5) we get

$$\beta d\left((\overline{Y}, \overline{K}), (\overline{Y}', \overline{K}')\right) \leq (\alpha + C\gamma)\mathbb{E}\int_0^T (e^{\beta s} - 1)\|\widehat{Y}_s\|ds$$

$$+ \lambda\mathbb{E}\int_0^T (e^{\beta s} - 1)\|D_s - D_s'\|ds. \tag{2.18}$$

Choose β large enough that $\frac{1}{\beta}(\alpha + C\gamma) \leq \lambda$. Then we get using (2.11) and (2.18)

$$d\left((\overline{Y}, \overline{K}), (\overline{Y}', \overline{K}')\right) \leq \lambda d\left((Y, K), (Y', K')\right). \tag{2.19}$$

Hence as $\lambda < 1$, the mapping ϕ is a strict contraction establishing thus the existence and uniqueness of the solution.   □

Let $\{(Y_t, Z_t, K_t)\ 0 \leq t \leq 1\}$ and $\{(Y_t', Z_t', K_t')\ 0 \leq t \leq 1\}$ denote two solutions of our RBSDE. For every $t \geq 0$ define

$$(\Delta Y_t, \Delta Z_t, \Delta K_t, \Delta D_t) = (Y_t - Y_t', Z_t - Z_t', K_t - K_t', D_t - D_t').$$

Then applying Ito's formula

$$\mathbb{E}|\Delta(n_i.Y_t)|^2 + \mathbb{E}\int_0^T |\Delta(n_i.Z_s)|^2 ds$$

$$= 2\mathbb{E}\int_0^T |\Delta(n_i.Y_s)| \left|n_i.f(s, \overline{Y}_s) - n_i.f(s, \overline{Y}_s')\right| ds$$

$$+ 2\mathbb{E}\int_0^T |\Delta(n_i.Y_s)| \left|D_i(s) - D_i'(s)\right| ds$$

$$+ 2\mathbb{E}\int_0^T |\Delta(n_i.Y_s)| \left|\sum_{k=1,k\neq i}^N n_i.v_k(s, \overline{Y}_s)\overline{D}_s^k - \sum_{k=1,k\neq i}^N n_i.v_k(s, \overline{Y}_t')\overline{D}_s'^k\right| ds.$$

Multiplying by a_i and adding leads to the existence of C such that

$$\mathbb{E}\left(\sum_{i=1}^{N} a_i |\Delta(n_i.Y_t)|^2\right) + \mathbb{E}\left(\int_0^T \sum_{i=1}^{N} a_i |\Delta(n_i.Z_s)|^2 ds\right)$$

$$\leq C\mathbb{E}\left(\int_0^T \sum_{i=1}^{N} a_i |\Delta(n_i.Y_s)| ds\right).$$

Therefore

$$Z = Z'.$$

References

1. Chaleyat-Maurel, El Karoui, N. and Marchal, B. (1980) Reflexion discontinue et systémes stochastiques. Ann. Proba., 8, 1049–1067.
2. El Karoui N, Kapoudjian C, Pardoux E, Peng S and Quenez M C. (1997) Reflected solutions of backward SDE's, and related obstacle problems for PDE's. Ann. Probab. 25 702–737.
3. Gegout-Petit, A. and Pardoux, E. (1996) Equations différentielles stochastiques rétrogrades refléchies dans un convexe. Stochastics and Stochastics Reports, 57, 111–128.
4. Ikeda, N. and Watanabe, S. (1989) Stochastic Differential Equations and Diffusion Processes. North-Holland, Amsterdam.
5. Liptser, R.Sh. and Shiryayev, A.N. (1989) Theory of Martingales. Kluwer Acad. Publ.
6. Lions P.L. and Sznitman A.S. (1984) Stochastic differential equations with reflecting boundary conditions. Comm. Pure Appl. Math., 37, 511–537.
7. Menaldi, J.L. and Robin, M. (1985) Reflected diffusion processes with jumps. Annals of Probability, 13:2, 319–341.
8. Ouknine, Y. (1998) Reflected backward stochastic differential equations with jumps. Stochastics and Stochastics Reports 65, 111–125.
9. Shashiashvili M. (1996) The Skorokhod oblique reflection problem in a convex polyhedron. Georgian Math. J. 3 153–176.
10. Tanaka, H. (1979) Stochastic differential equations with reflecting boundary condition in convex regions. Hiroshima Math. J. 9, 163–177.
11. Watanabe, S. (1971) On stochastic differential equations for multidimensional diffusion processes with boundary conditions. J. Math. Kyoto Univ. 11, 169–180.

Chapter 3
On the Energy Decay of a Nonhomogeneous Hybrid System of Elasticity

Moulay Driss Aouragh and Abderrahman El Boukili

Abstract In this paper, we study the boundary stabilizing feedback control problem of well-known Scole model that has nonhomogeneous spatial parameters. By using an abstract result of Riesz basis, we show that the closed-loop system is a Riesz spectral system. The asymptotic distribution of eigenvalues, the spectrum-determinded growth condition and the exponential stability are concluded.

Keywords Euler-Bernoulli beam · Boundary control · Stabilization · Riesz basis

3.1 Introduction

The boundary and internal control problem of flexible structure has recently attracted much attention with the rapid development of high technology such as space science and flexible robots. In this paper, we study the boundary feedback stabilization of the nonuniform Scole model. Consisting of an elastic beam, linked to a rigid antenna, this dynamical system is governed by the nonuniform Euler–Bernoulli equation for the vibration of the elastic beam and the Newton–Euler rigid body equation for the oscillation of the antenna. The nonuniform Scole model in the case of a hinged (or "pinned") beam, correspond to the following hybrid system:

M. D. Aouragh (✉)
MAMCS Group, Department of Maths, M2I Laboratory, FST, Moulay Ismaïl University, Errachidia, Morocco

A. El Boukili
Department of Physics, FST, Moulay Ismaïl University, Errachidia, Morocco

© Springer Nature Switzerland AG 2020
S. Dos Santos et al. (eds.), *Recent Advances in Mathematics and Technology*,
Applied and Numerical Harmonic Analysis,
https://doi.org/10.1007/978-3-030-35202-8_3

$$\begin{cases} \rho(x)y_{tt}(x,t) + (EI(x)y_{xx}(x,t))_{xx} = 0\,, & 0 < x < 1, t > 0, \\ y_x(0,t) = 0, (y + (EI(.)y_{xx})_x + ay_t)(0,t) = 0, & t > 0, \\ (my_{tt} - (EI(.)y_{xx})_x)(1,t) = -by_t(1,t) & t > 0, \\ (Jy_{xtt} + EI(.)y_{xx})(1,t) = -cy_{xt}(1,t), & t > 0, \\ y(x,0) = y_0(x)\,, \quad y_t(x,0) = y_1(x)\,, & 0 < x < 1, \end{cases}$$

(3.1)

where y represents the transversal displacement of the beam, x denotes the position, and t denotes the time. $\rho(x)$ is the mass density of the beam and $EI(x)$ is its flexural rigidity. m is the mass of the antenna and J is its moment of inertia. a, b, and c, are constants feedback gains.

For further description of the physical structure of the system, we refer to Littman–Markus [5]. Furthermore, the coefficients are supposed to be variable because it is common in engineering, to adopt problems with nonhomogeneous materials such as smart materials [4]. Notice that the boundary feedbacks can be realized by means of passive mechanical systems of springs-dampers similar to those used in [1]. The stabilization problem of system (3.1) has been the subject of many studies. When the coefficients ρ, EI are supposed to be constants, Rao in [9] establish the uniform energy decay by using energy multiplier method [6]. It seems to be difficult to extend this method to the nonuniform case. In this paper, we extend the results obtained in [9] to variable coefficients. By using the Riesz basis approach, we show that the generalized eigenfunctions of the system form a Riesz basis for the state Hilbert space. As a consequence, the asymptotic expressions of eigenvalues together with exponential stability are obtained.

The rest of this paper is organized as follows. In Sect. 3.2, the well-posedness and the asymptotic stability of the closed-loop system are established. Section 3.3 is devoted to the asymptotic analysis for the eigenpairs of the closed-loop system. Finally, in Sect. 3.4, we prove the Riesz basis property, the spectrum determined growth condition and the optimal decay rate.

Throughout this paper, we assume that

$$(EI(.), \rho(.)) \in [C^4(0,1)]^2, \ \rho, \ EI > 0, \ m, \ J > 0, \tag{3.2}$$

and the constants a, b, and c satisfy the dissipation condition

$$a > 0, \ b \geq 0, \ c > 0. \tag{3.3}$$

3.2 Well-Posedness and Asymptotic Stability

We consider system (3.1) on the following complex Hilbert space:

$$\mathbb{H} = \mathbb{V} \times L^2(0,1) \times \mathbb{C}^2, \tag{3.4}$$

where

$$\mathbb{V} = \{f \in H^2(0, 1) / f'(0) = 0\}, \tag{3.5}$$

equipped with the inner product defined as
$\forall(F = (f_1, g_1, \zeta_1, \delta_1), G = (f_2, g_2, \zeta_2, \delta_2)) \in \mathbb{H}^2$

$$(F, G)_{\mathbb{H}} = \int_0^1 (\rho(x)g_1(x)\overline{g_2(x)} + EI(x)f_1''(x)\overline{f_2''(x)})dx + f_1(0)\overline{f_2(0)}$$

$$+ \frac{1}{m}\zeta_1\overline{\zeta_2} + \frac{1}{J}\delta_1\overline{\delta_2}. \tag{3.6}$$

Then, we define an operator as follows: $\mathbb{A} : D(\mathbb{A}) \subset \mathbb{H} \rightarrow \mathbb{H}$

$$\begin{cases} D(\mathbb{A}) = \{(f, g, \zeta, \delta) \in (H^4(0, 1) \cap \mathbb{V}) \times \mathbb{V} \times \mathbb{C}^2 / f(0) + (EI(.)f'')'(0) + ag(0) = 0, \\ \qquad \zeta = mg(1), \delta = Jg'(1)\} \\ \mathbb{A}(f, g, \zeta, \delta) = \left(g, \dfrac{-(EI(.)f''(.))''}{\rho(.)}, (EI(.)f'')'(1) - bg(1), -(EI(1)f''(1) + cg'(1)) \right), \end{cases} \tag{3.7}$$

with the initial condition $Y_0 = \left(y_0, y_1, my_1(1), Jy_1'(1)\right)$, the system (3.1) can be written as an evolutionary equation in $\mathbb{H}$:

$$\begin{cases} \dfrac{dY(t)}{dt} = \mathbb{A}Y(t), \\ Y(t) = (y(., t), y_t(., t), my_t(1, t), Jy_{xt}(1, t)), \ Y(0) = Y_0. \end{cases} \tag{3.8}$$

We have the following Lemma

Lemma 3.1 *Let the operator $\mathbb{A}$ defined by (3.7). Then $\mathbb{A}$ is a densely defined, closed dissipative operator in $\mathbb{H}$, and $\mathbb{A}^{-1}$ exists and is compact on $\mathbb{H}$. Moreover, $\mathbb{A}$ generates a C_0 semigroup of contractions $e^{\mathbb{A}t}$ on $\mathbb{H}$ and the spectrum $\sigma(\mathbb{A})$ of $\mathbb{A}$ consists only of the isolated eigenvalues.*

Proof Let $(f, g, \zeta, \delta) \in D(\mathbb{A})$, then we have

$$Re(\mathbb{A}Y, Y)_{\mathbb{H}} = -a|g(0)|^2 - b|g(1)|^2 - c|g'(1)|^2. \tag{3.9}$$

Thus $\mathbb{A}$ is dissipative in $\mathbb{H}$. Next, we show that $\mathbb{A}^{-1}$ exists. Let $(u, v, \omega, \xi) \in \mathbb{H}$, we will find $(f, g, \zeta, \delta) \in D(\mathbb{A})$ such that

$$\mathbb{A}(f, g, \zeta, \delta) = (u, v, \omega, \xi) \in \mathbb{H},$$

which yields

$$
\begin{cases}
g = u, \zeta = mg(1) = mu(1), \delta = Jg'(1) = Ju'(1), \\
(EI(.)f'')''(x) = -\rho(x)v(x), \\
f'(0) = 0, \ f(0) + (EI(.)f'')'(0) + au(0) = 0, \\
(EI(.)f'')'(1) - bu(1) = \omega, \\
-(EI(1)f''(1) + cu'(1)) = \xi.
\end{cases}
$$

After a simple calculation, we show that

$$
f(x) = f(0) - \int_0^x \int_0^y dr dy \left[\frac{\beta(1-r) + \alpha}{EI(r)} + \frac{1}{EI(r)} \int_r^1 \int_s^1 \rho(x)v(x)dt ds \right],
$$

where

$$
\begin{cases}
f(0) = -(\beta + au(0) + \int_0^1 \int_r^1 \rho(x)v(x)ds dr), \\
\alpha = \xi + cu'(1), \ \beta = \omega + bu(1).
\end{cases}
$$

Thus, $\mathbb{A}^{-1}$ exists and is bounded in $\mathbb{H}$. Furthermore, the Sobolev embedding theorem, implies that $\mathbb{A}^{-1}$ is compact on $\mathbb{H}$ and the Lumer–Phillips theorem [8] can be applied to conclude that $\mathbb{A}$ generates a C_0 semigroup of contractions $e^{\mathbb{A}t}$ in $\mathbb{H}$. The Lemma is proved. $\square$

Now, we turn our attention to the asymptotic stability of the system.

Lemma 3.2 *Let $\mathbb{A}$ be the operator defined by (3.7). Then $\Re e(\mathbb{A}) < 0$ and hence the system (3.1) is asymptotically stable.*

Proof It suffices to show that $\{i\gamma, \gamma \in \mathbb{R}\} \subset \rho(\mathbb{A})$. Assume that this is false. This together with Lemma 3.1 implies that there exists nonzero $\gamma \in \mathbb{R}$ such that $i\gamma \in \sigma(\mathbb{A})$, where $\sigma(\mathbb{A})$ is the point spectrum, i.e., there exists $\phi = (f, g, \zeta, \delta) \in D(\mathbb{A})$ satisfying without loss of generality, the conditions $\|\phi\|_{\mathbb{H}} = 1$ and $(i\gamma - \mathbb{A})\phi = 0$ i.e.,

$$
\begin{cases}
(EI(.)f'')''(x) - \gamma^2 \rho(x)f(x) = 0, \\
f'(0) = 0, \ -(EI(.)f'')'(0) = (1 + i\gamma a)f(0), \\
(EI(.)f'')'(1) = (-\gamma^2 m + i\gamma b)f(1), \\
-EI(1)f''(1) = (-\gamma^2 J + i\gamma c)f'(1), \\
g = i\gamma f, \zeta = i\gamma mf(1), \delta = i\gamma Jf'(1).
\end{cases} \tag{3.10}
$$

Using (3.9), we obtain $g'(1) = f'(1) = 0$ and $f(0) = 0$, which further implies by means of (3.10) that $f''(1) = 0$ and the system (3.10) yields

$$
\begin{cases}
(EI(.)f'')''(x) - \gamma^2 \rho(x) f(x) = 0, \\
f(0) = f'(0) = (EI(.)f'')'(0) = 0, \\
f'(1) = f''(1) = 0, \\
(EI(.)f'')'(1) = (-\gamma^2 m + i\gamma b) f(1).
\end{cases}
\tag{3.11}
$$

1. If $b > 0$, then from (3.9), $g(1) = f(1) = 0$, by means of (3.11), we have

$$
(EI(.)f'')'(1) = 0
$$

and the system (3.11) yields

$$
\begin{cases}
(EI(.)f'')''(x) - \gamma^2 \rho(x) f(x) = 0, \\
f(0) = f'(0) = (EI(.)f'')'(0) = 0, \\
f(1) = (EI(.)f'')'(1) = 0, \\
EI(1) f''(1) = 0.
\end{cases}
\tag{3.12}
$$

It has been proved in [3] that the above system has only the trivial solution, i.e., $f = 0$. Then $\phi = 0$, which contradict the first that $\|\phi\|_{\mathbb{H}} = 1$.

2. If $b = 0$. First, assume that

$$
f(1) > 0 \text{ (the negative case is similar)},
$$

which implies by the last boundary condition in (3.11) that

$$
(EI f'')'(1) < 0.
$$

Let $[c, 1]$ be a subspace of $[0, 1]$ so that $f(x) > 0$ for each $x \in (c, 1]$, $f(c) = 0$. Then,

$$
(EI(.)f'')''(x) > 0, \text{ for any } x \in (c, 1].
$$

Hence, $(EI(.)f'')'$ is increasing in $(c, 1]$. Since

$$
(EI(.)f'')'(1) < 0,
$$

we have

$$
(EI(.)f'')'(x) < 0, \text{ for any } x \in (c, 1].
$$

It follows that $EI(x)f''(x)$ is decreasing in $(c, 1]$. Since

$$EI(1)f''(1) = 0,$$

we have

$$f''(x) > 0, \text{ for any } x \in (c, 1).$$

So, $f'(x)$ is increasing in $(c, 1)$. Since $f'(1) = 0$, we have

$$f'(x) < 0, \text{ for any } x \in (c, 1).$$

Hence, $f(x)$ is decreasing in $(c, 1)$, and so,

$$f(c) > f(1) > 0,$$

contradicts the assumption that $f(c) = 0$. Therefore, $f(1) = 0$. Now, (3.11) implies that f satisfies system (3.12). We can conclude as in 1. The Lemma 3.2 (in the end of proof of Lemma 3.2) is proved. □

3.3 Asymptotic Expressions of Eigenfrequencies

Note that

$$\mathbb{A}\phi = \lambda\phi, \phi = (f, g, \zeta, \delta), \tag{3.13}$$

yields

$$\begin{cases} (EI(.)f'')''(x) + \lambda^2 \rho(x)f(x) = 0, \ 0 < x < 1, \\ f'(0) = 0, \ f(0) + (EI(.)f'')'(0) + ag(0) = 0, \\ (EI(.)f'')'(1) = (\lambda^2 m + \lambda b)f(1), \\ -EI(1)f''(1) = (\lambda^2 J + \lambda c)f'(1), \\ g(x) = \lambda f(x), \zeta = mg(1), \delta = Jg'(1). \end{cases} \tag{3.14}$$

Writing (3.14) in the standard form of a linear differential operator with homogeneous boundary conditions, we obtain

$$\begin{cases} f^{(4)}(x) + \dfrac{2EI'(x)}{EI(x)} f'''(x) + \dfrac{EI''(x)}{EI(x)} f''(x) + \lambda^2 \dfrac{EI(x)}{\rho(x)} f(x) = 0, \ 0 < x < 1, \\[2mm] f'(0) = 0, \ \lambda f(0) + a_1 f'''(0) + a_2 f''(0) + a_3 f(0) = 0, \\[2mm] \lambda^2 f(1) + a_4 \lambda f(1) - a_5 f'''(1) - a_6 f''(1) = 0, \\[2mm] \lambda^2 f'(1) + a_7 \lambda f'(1) + a_8 f''(1) = 0, \end{cases}$$

$$(3.15)$$

where

$$\begin{cases} a_1 = \dfrac{EI(0)}{a}, \ a_2 = \dfrac{EI'(0)}{a}, \ a_3 = \dfrac{1}{a}, a_4 = \dfrac{b}{m}, \\[3mm] a_5 = \dfrac{EI(1)}{m}, \ a_6 = \dfrac{EI'(1)}{m}, \ a_7 = \dfrac{c}{m}, \ a_8 = \dfrac{EI(1)}{J}. \end{cases}$$

$$(3.16)$$

In order to simplify the computations, we introduce a spatial scale transformation in x.

$$\Phi(z) = f(x), \ z = z(x) = \frac{1}{p} \int_0^x \left(\frac{\rho(s)}{EI(s)} \right)^{1/4} ds, \ p = \int_0^1 \left(\frac{\rho(s)}{EI(s)} \right)^{1/4} ds,$$

$$(3.17)$$

then Φ satisfies the following system:

$$\begin{cases} \Phi^{(4)}(z) + a(z)\Phi'''(z) + b(z)\Phi''(z) + c(z)\Phi'(z) + \lambda^2 p^4 \Phi(z) = 0, \\[2mm] \Phi'(0) = 0, \ \lambda\Phi(0) + b_1 \Phi'''(0) + b_2 \Phi''(0) + b_3 \Phi'(0) + a_3 \Phi(0) = 0, \\[2mm] \lambda^2 \Phi(1) + a_4 \lambda \Phi(1) - b_4 \Phi'''(1) - b_5 \Phi''(1) - b_6 \Phi'(1) = 0, \\[2mm] \lambda^2 \Phi'(1) + a_7 \lambda \Phi'(1) + b_7 \Phi''(1) + b_8 \Phi'(1) = 0, \end{cases}$$

$$(3.18)$$

where $a(z)$, $b(z)$, and $c(z)$ are the smooth functions defined by

$$\begin{cases} a(z) = \dfrac{6z''}{z'^2} + \dfrac{2EI'(x)}{z'EI(x)}, \\[3mm] b(z) = \dfrac{3z''^2}{z'^4} + \dfrac{6z''EI'(x)}{z'^3 EI(x)} + \dfrac{EI''(x)}{z'^2 EI(x)} + \dfrac{4z'''}{z'^3}, \\[3mm] c(z) = \dfrac{z''''}{z'^4} + \dfrac{2z'''EI'(x)}{z'^4 EI(x)} + \dfrac{z''EI''(x)}{z'^4 EI(x)}, \end{cases}$$

$$(3.19)$$

and

$$\begin{cases} b_1 = a_1 z'^3(0), \ b_2 = 3a_1 z'(0)z''(0) + a_2 z'^2(0), \\ b_3 = a_1 z'''(0) + a_2 z''(0), \ b_4 = a_5 z'^3(1), \\ b_5 = 3a_5 z'(1)z''(1) + a_6 z'^2(1), \ b_6 = a_5 z'^3(1) + a_6 z''(1), \\ b_7 = a_8 z'(1), \ b_8 = \dfrac{a_8 z''(1)}{z'(1)}. \end{cases} \tag{3.20}$$

Equation (3.18) can be simplified by applying another invertible transformation

$$\varphi(z) = e^{1/4 \int_0^z a(s)ds} \, \Phi(z), \tag{3.21}$$

which allows one to cancel the term $a(z)\Phi'''(z)$ in (3.18); hence, φ satisfies the following equivalent eigenvalue problem:

$$\begin{cases} \varphi^{(4)}(z) + a_1(z)\varphi''(z) + a_2(z)\varphi'(z) + a_3(z)\varphi(z) + \lambda^2 p^4 \varphi(z) = 0, \\ \varphi'(0) - \dfrac{a(0)}{4}\varphi(0) = 0, \ \lambda\varphi(0) + b_1\varphi'''(0) + F_1(\varphi(0), \varphi'(0), \varphi''(0)) = 0, \\ \lambda^2\varphi(1) + a_4\lambda\varphi(1) - b_4\varphi'''(1) + F_2(\varphi(1), \varphi'(1), \varphi''(1)) = 0, \\ \lambda^2(\varphi'(1) - \dfrac{a(1)}{4}\varphi(1)) + a_7\lambda(\varphi'(1) - \dfrac{a(1)}{4}\varphi(1)) + F_3(\varphi(1), \varphi'(1), \varphi''(1)) = 0, \end{cases} \tag{3.22}$$

where $a_1(z)$, $a_2(z)$ and $a_3(z)$ are the smooth functions defined by

$$\begin{cases} a_1(z) = -\dfrac{3a'(z)}{2} - \dfrac{3a^2(z)}{8} + b(z), \\ a_2(z) = \dfrac{a^3(z)}{8} - a''(z) - \dfrac{a(z)b(z)}{2} + c(z), \\ a_3(z) = \dfrac{3a'^2(z)}{16} - \dfrac{a'''(z)}{4} - \dfrac{3a^4(z)}{256} + \dfrac{3a^2(z)a'(z)}{32} \\ \qquad\quad + b(z)(\dfrac{a^2(z)}{16} - \dfrac{a'(z)}{4}) - \dfrac{a(z)c(z)}{4}, \end{cases} \tag{3.23}$$

and $F_1(x_1, x_2, x_3)$, $F_2(x_1, x_2, x_3)$, and $F_3(x_1, x_2, x_3)$ are linear combinations of $x_1, x_2,$ and x_3.

To estimate asymptotically the solutions to the eigenvalue problem (3.22), we proceed as in [7]. First due to Lemma 3.2 and the fact that eigenvalues of $\mathbb{A}$ are symmetric with respect to the real axis, we only need to consider those $\lambda \in \sigma(\mathbb{A})$ that satisfy $\dfrac{\pi}{2} \leq \arg\lambda \leq \pi$, which we assume in the sequel. Next, we set $\lambda = \tau^2$ and hence

$$\frac{\pi}{4} \le \arg \tau \le \frac{\pi}{2}.$$

Now, let us choose ω_j, $j = 1, 2, 3, 4$ as follows:

$$\omega_1 = \frac{-1+i}{\sqrt{2}}, \quad \omega_2 = \frac{1+i}{\sqrt{2}}, \quad \omega_3 = -\omega_2, \quad \omega_4 = -\omega_1,$$

consequently, we have for $\tau \in S = \left\{ \tau / \frac{\pi}{4} \le \arg \tau \le \frac{\pi}{2} \right\}$

$$\begin{cases} Re(\tau\omega_1) = -\mid \tau \mid \sin(\arg \tau + \frac{\pi}{4}) \le -\frac{\sqrt{2} \mid \tau \mid}{2} < 0, \\[2mm] Re(\tau\omega_2) = \mid \tau \mid \cos(\arg \tau + \frac{\pi}{4}) \le 0. \end{cases} \tag{3.24}$$

In order to analyze the asymptotic distribution of eigenpairs for (3.22), we need the following result [10].

Lemma 3.3 *For $\mid \tau \mid$ large enough and $\tau \in S$, there are four linearly independent asymptotic fundamental solutions φ_j , $j = 1, 2, 3, 4$, to*

$$\varphi^{(4)}(z) + a_1(z)\varphi''(z) + a_2(z)\varphi'(z) + a_3(z)\varphi(z) + \tau^4 p^4 \varphi(z) = 0, \tag{3.25}$$

such that

$$\begin{cases} \varphi_j(z, \tau) = e^{\tau\omega_j z} \left(1 + \frac{\varphi_{j1}(z)}{\tau} + O\left(\tau^{-2}\right) \right), \\[3mm] \varphi_j'(z, \tau) = \tau\omega_j e^{\tau\omega_j z} \left(1 + \frac{\varphi_{j1}(z)}{\tau} + O\left(\tau^{-2}\right) \right), \\[3mm] \varphi_j''(z, \tau) = (\tau\omega_j)^2 e^{\tau\omega_j z} \left(1 + \frac{\varphi_{j1}(z)}{\tau} + O\left(\tau^{-2}\right) \right), \\[3mm] \varphi_j'''(z, \tau) = \left(\tau\omega_j\right)^3 e^{\tau\omega_j z} \left(1 + \frac{\varphi_{j1}(z)}{\tau} + O\left(\tau^{-2}\right) \right), \end{cases} \tag{3.26}$$

where

$$\varphi_{j1}(z) = -\frac{1}{4\omega_j} \int_0^z a_1(s)ds.$$

Hence, for $j = 1, 2, 3, 4$,

$$\varphi_{j1}(0) = 0, \quad \varphi_{j1}(1) = -\frac{1}{4\omega_j} \int_0^1 a_1(s)ds = \frac{1}{\omega_j}\mu, \quad \mu = -\frac{1}{4} \int_0^1 a_1(s)ds.$$

For convenience, we introduce the notation $[r]_j = r + O\left(\tau^{-j}\right)$ for $j = 1, 2$. From Lemma 3.3, one can write the asymptotic solution of (3.22) as follows:

$$\varphi(z) = d_1\varphi_1(z) + d_2\varphi_2(z) + d_3\varphi_3(z) + d_4\varphi_4(z), \tag{3.27}$$

where φ_j, $j = 1, 2, 3, 4$ are defined by Lemma 3.3 and d_j, $j = 1, 2, 3, 4$ are chosen so that φ satisfy the boundary conditions of (3.22). Note that $\lambda = \tau^2 \neq 0$, is the eigenvalue of (3.22) if and only if τ satisfies the characteristic determinant

$$\Delta(\tau) = \begin{vmatrix} \tau\omega_1[1 - \frac{a_0}{\tau\omega_1}]_2 & \tau\omega_2[1 - \frac{a_0}{\tau\omega_2}]_2 \\ \tau^3\omega_1^3[b_1 - \frac{\omega_1}{\tau}]_2 & \tau^3\omega_2^3[b_1 - \frac{\omega_2}{\tau}]_2 \\ \tau^4[1 + \frac{\mu + b_4}{\tau\omega_1}]_2 e^{\tau\omega_1} & \tau^4[1 + \frac{\mu + b_4}{\tau\omega_2}]_2 e^{\tau\omega_2} \\ \tau^5[\omega_1 + \frac{b_9}{\tau}]_2 e^{\tau\omega_1} & \tau^5[\omega_2 + \frac{b_9}{\tau}]_2 e^{\tau\omega_2} \end{vmatrix} \tag{3.28}$$

$$\begin{matrix} \tau\omega_3[1 - \frac{a_0}{\tau\omega_3}]_2 \\ \tau^4\omega_3^3[b_1 - \frac{\omega_3}{\tau}]_2 \\ \tau^4[1 + \frac{\mu + b_4}{\tau\omega_3}]_2 e^{\tau\omega_3} \\ \tau^5[\omega_3 + \frac{b_9}{\tau}]_2 e^{\tau\omega_3} \end{matrix}$$

$$\begin{matrix} \tau\omega_4[1 - \frac{a_0}{\tau\omega_4}]_2 \\ \tau^3\omega_4^3[b_1 - \frac{\omega_4}{\tau}]_2 \\ \tau^4[1 + \frac{\mu + b_4}{\tau\omega_4}]_2 e^{\tau\omega_4} \\ \tau^5[\omega_4 + \frac{b_9}{\tau}]_2 e^{\tau\omega_4} \end{matrix} \Bigg|,$$

where

$$a_0 = \frac{a(0)}{4}, \quad b_9 = \mu - \frac{a(1)}{4}.$$

Noting that from (3.24)

$$|e^{\tau\omega_2}| \leq 1, |e^{\tau\omega_1}| = O(e^{-q|\tau|}) \text{ as } |\tau| \to +\infty,$$

for some constant $q > 0$, then each element of the matrix in (3.28) is bounded, we may rewrite (3.28) as

$$\tau^{-13} e^{\tau(\omega_1+\omega_2)} \Delta(\tau) = \begin{vmatrix} \omega_1(1 - \frac{a_0}{\tau\omega_1}) & \omega_2(1 - \frac{a_0}{\tau\omega_2}) & -\omega_2(1 + \frac{a_0}{\tau\omega_2})e^{\tau\omega_2} \\ \omega_1^3(b_1 - \frac{\omega_1}{\tau}) & \omega_2^3(b_1 - \frac{\omega_2}{\tau}) & -\omega_2^3(b_1 + \frac{\omega_2}{\tau})e^{\tau\omega_2} \\ 0 & (1 + \frac{\mu+b_4}{\tau\omega_2})e^{\tau\omega_2} & (1 - \frac{\mu+b_4}{\tau\omega_2}) \\ 0 & (\omega_2 + \frac{b_9}{\tau})e^{\tau\omega_2} & (-\omega_2 + \frac{b_9}{\tau}) \end{vmatrix}$$

$$\left. \begin{matrix} 0 \\ 0 \\ (1 - \frac{\mu+b_4}{\tau\omega_1}) \\ (-\omega_1 + \frac{b_9}{\tau}) \end{matrix} \right| + O\left(\tau^{-2}\right). \tag{3.29}$$

A direct calculation gives

$$\tau^{-13} e^{\tau(\omega_1+\omega_2)} \Delta(\tau)$$

$$= b_1(\omega_1\omega_2^3 - \omega_1^3\omega_2)\left[\omega_2 - \omega_1 + (\mu + b_4)\left(\omega_1\omega_2^{-1} - \omega_2\omega_1^{-1}\right)\tau^{-1}\right]$$

$$+ (\omega_2 - \omega_1)(\omega_1^3 - \omega_2^3)(\omega_1\omega_2 + a_0 b_1)\tau^{-1} + e^{2\tau\omega_2}\Big\{ b_1(\omega_1^3\omega_2 - \omega_1\omega_2^3)$$

$$\times \left[\omega_1 + \omega_2 + (\mu + b_4)(\omega_2\omega_1^{-1} - \omega_1\omega_2^{-1})\tau^{-1}\right] + (\omega_1 + \omega_2)(\omega_1^3 + \omega_2^3)$$

$$\times (b_1 a_0 - \omega_1\omega_2)\tau^{-1}\Big\} + O\left(\tau^{-2}\right).$$

A straightforward simplification will arrive at the following result.

Theorem 3.1 *Let $\lambda = \tau^2$ where $\tau \in S$.*

1. *The characteristic determinant $\Delta(\tau)$ of the eigenfunction problem (3.22) has the following asymptotic expression in the sector S*

$$\tau^{-13} e^{\tau(\omega_1+\omega_2)} \Delta(\tau) = 2\left[-\sqrt{2}i b_1 + (b_{10} - 1)\tau^{-1}\right]$$

$$+ 2e^{2\tau\omega_2}\left[-\sqrt{2}b_1 + (-b_{10} - 1)\tau^{-1}\right] + O\left(\tau^{-2}\right), \tag{3.30}$$

   where $b_{10} = b_1[2(\mu + b_4) + a_0]$.

2. *Let $\sigma(\mathbb{A}) = \{\lambda_n, \overline{\lambda}_n, n \in \mathbb{N}\}$, be the eigenvalues of $\mathbb{A}$, then for $k = n - \frac{1}{4}$ and $\tau_n \in S$, the following asymptotic expression holds*

$$\tau_n = \frac{1}{\omega_2} k\pi i - \frac{1}{2\sqrt{2}b_1} \left[\frac{(1-i)b_{10} + (1+i)}{k\pi i} \right] + O\left(n^{-2}\right)$$

$$\lambda_n = -\frac{1}{b_1} + (k\pi)^2 i + \frac{b_{10}i}{b_1} + O\left(n^{-1}\right),$$

(3.31)

for sufficiently large positive integer n. Moreover, by (3.16) and (3.20), we obtain

$$\lim_{n \to +\infty} Re(\lambda_n) = -\frac{1}{b_1} = -\frac{a}{EI(0)z'^3(0)} < 0.$$

(3.32)

3. λ_n is geometrically simple when n is large enough.

Proof Note that $\lambda = \tau^2 \in \sigma(\mathbb{A})$, where $\tau \in S$ if and only if

$$- \sqrt{2}ib_1 e^{-\tau\omega_2} - \sqrt{2}b_1 e^{\tau\omega_2} + \frac{(b_{10} - 1)}{\tau} e^{-\tau\omega_2} - \frac{(b_{10} + 1)}{\tau} e^{\tau\omega_2} + O\left(\tau^{-2}\right) = 0,$$

(3.33)

which can be written as

$$- \sqrt{2}ib_1 e^{-\tau\omega_2} - \sqrt{2}b_1 e^{\tau\omega_2} + O\left(\tau^{-1}\right) = 0.$$

(3.34)

Obviously, the equation

$$ie^{-\tau\omega_2} + e^{\tau\omega_2} = 0,$$

(3.35)

has solutions

$$\widetilde{\tau}_n = \frac{1}{\omega_2} k\pi i, \; n \in \mathbb{N}, \quad k = n - \frac{1}{4}.$$

(3.36)

Applying Rouche's theorem to (3.34), we obtain

$$\tau_n = \widetilde{\tau}_n + \alpha_n = \frac{1}{\omega_2} k\pi i + \alpha_n, \alpha_n = O\left(n^{-1}\right), n = N, N+1 \ldots,$$

(3.37)

where N is large positive integer.
Substituting τ_n into (3.33) and using the fact that $e^{\widetilde{\tau}_n\omega_2} = -ie^{-\widetilde{\tau}_n\omega_2}$, we obtain

$$- \sqrt{2}ib_1 e^{-\alpha_n\omega_2} + \sqrt{2}ib_1 e^{\alpha_n\omega_2} + (b_{10} - 1)\widetilde{\tau}_n^{-1} e^{-\alpha_n\omega_2} + i(b_{10} + 1)\widetilde{\tau}_n^{-1} e^{\alpha_n\omega_2}$$
$$+ O\left(\widetilde{\tau}_n^{-2}\right) = 0.$$

On the other hand, expanding the exponential function according to its Taylor series, we obtain

$$\alpha_n = -\frac{1}{2\sqrt{2}b_1} \left[\frac{b_{10}(1-i) + (1+i)}{k\pi i} \right] + O\left(n^{-2}\right).$$

(3.38)

Substituting this estimate in (3.37), we have,

$$\tau_n = \frac{k\pi i}{\omega_2} - \frac{1}{2\sqrt{2}b_1}\left[\frac{b_{10}(1-i)+(1+i)}{k\pi i}\right] + O\left(n^{-2}\right), n = N, N+1, \ldots.$$

(3.39)

Finally, recall that $\lambda_n = \tau_n^2$, $\omega_2 = \frac{1+i}{\sqrt{2}}$, $\omega_2^2 = i$, and hence the last estimate yields

$$\lambda_n = -\frac{1}{b_1} + (k\pi)^2 i + \frac{b_{10}i}{b_1} + O\left(n^{-1}\right), n = N, N+1, \ldots,$$

where N is sufficiently large.

Since the matrix in (3.29) has rank 3 for each sufficiently large n, there is only one linearly independent solution φ_n to (3.22) for $\tau = \tau_n$. Hence, each λ_n is geometrically simple for n sufficiently large. The theorem is proved. □

Theorem 3.2 *Let $\lambda_n = \tau_n^2$ where $\tau_n \in S$ is given by (3.31). Then the corresponding eigenfunction $\{\phi_n = (f_n, \lambda_n f_n, \zeta_n, \delta_n), \overline{\phi}_n = (\overline{f}_n, \overline{\lambda}_n \overline{f}_n, \overline{\zeta}_n, \overline{\delta}_n)\}$ has the following asymptotic:*

$$\lambda_n f_n(x) = e^{-1/4\int_0^z a(s)ds}\left[\sqrt{2}cos(n-1/4)\pi z - (-1)^n e^{-(n-1/4)\pi(1-z)} + O\left(n^{-1}\right)\right],$$

(3.40)

$$f_n''(x) = \frac{1}{p^2}\left(\frac{\rho(x)}{EI(x)}\right)^{1/2} e^{-1/4\int_0^z a(s)ds}\left[\sqrt{2}icos(n-1/4)\pi z + i(-1)^n\right.$$

$$\left. \times e^{-(n-1/4)\pi(1-z)} + O\left(n^{-1}\right)\right],$$

(3.41)

$$\zeta_n = O\left(n^{-1}\right), \delta_n = O\left(n^{-2}\right).$$

(3.42)

Proof From (3.25), (3.26), (3.28), and a simple fact of linear algebra, the eigenfunction φ_n corresponding to λ_n is given by

$$\varphi_n(z) = \begin{vmatrix} \tau_n\omega_1[1]_1 & \tau_n\omega_2[1]_1 & \tau_n\omega_3[1]_1 & \tau_n\omega_4[1]_1 \\ e^{\tau_n\omega_1 z}[1]_1 & e^{\tau_n\omega_2 z}[1]_1 & e^{\tau_n\omega_3 z}[1]_1 & e^{\tau_n\omega_4 z}[1]_1 \\ \tau_n^4 e^{\tau_n\omega_1}[1]_1 & \tau_n^4 e^{\tau_n\omega_2}[1]_1 & \tau_n^4 e^{\tau_n\omega_3}[1]_1 & \tau_n^4 e^{\tau_n\omega_4}[1]_1 \\ \tau_n^5\omega_1 e^{\tau_n\omega_1}[1]_1 & \tau_n^5\omega_2 e^{\tau_n\omega_2}[1]_1 & \tau_n^5\omega_3 e^{\tau_n\omega_3}[1]_1 & \tau_n^5\omega_4 e^{\tau_n\omega_4}[1]_1 \end{vmatrix},$$

(3.43)

then

$$
\omega_2^2 e^{\tau_n(\omega_1+\omega_2)}\varphi_n(z) = \tau_n^{10}
\begin{vmatrix}
-[1]_1 & i[1]_1 & -ie^{\tau_n\omega_2}[1]_1 & e^{\tau_n\omega_1}[1]_1 \\
e^{\tau_n\omega_1 z}[1]_1 & e^{\tau_n\omega_2 z}[1]_1 & e^{\tau_n\omega_2(1-z)}[1]_1 & e^{\tau_n\omega_1(1-z)}[1]_1 \\
e^{\tau_n\omega_1}[1]_1 & e^{\tau_n\omega_2}[1]_1 & [1]_1 & [1]_1 \\
-e^{\tau_n\omega_1}[1]_1 & ie^{\tau_n\omega_2}[1]_1 & -i[1]_1 & -[1]_1
\end{vmatrix}
$$

$$
= \tau_n^{10}
\begin{vmatrix}
-1 & i & -ie^{\tau_n\omega_2} & 0 \\
e^{\tau_n\omega_1 z} & e^{\tau_n\omega_2 z} & e^{\tau_n\omega_2(1-z)} & e^{\tau_n\omega_1(1-z)} \\
0 & e^{\tau_n\omega_2} & 1 & 1 \\
0 & ie^{\tau_n\omega_2} & -i & 1
\end{vmatrix}
+ O\left(\tau_n^{-1}\right)
$$

$$
= \tau_n^{10}\Big\{ -(1+i)e^{\tau_n\omega_2 z} + (1-i)e^{\tau_n\omega_2(1-z)}e^{\tau_n\omega_2} + 2ie^{\tau_n\omega_1(1-z)}e^{\tau_n\omega_2}
$$
$$
+ (1-i)e^{\tau_n\omega_1 z} - (i+1)e^{\tau_n\omega_1 z}e^{2\tau_n\omega_2}\Big\} + O\left(\tau_n^{-1}\right).
$$

It follows from (3.35) that $e^{2\tau_n\omega_2} = -i + O\left(\tau_n^{-1}\right)$, and hence the last estimate yields

$$
\frac{-\omega_2^2 e^{\tau_n(\omega_1+\omega_2)}\varphi_n(z)}{(1+i)\tau_n^{10}}
$$
$$
= \Big\{ e^{\tau_n\omega_2 z} + ie^{\tau_n\omega_2(1-z)}e^{\tau_n\omega_2} - (i+1)e^{\tau_n\omega_1(1-z)}e^{\tau_n\omega_2}\Big\} + O\left(n^{-1}\right).
$$

Similarly

$$
\frac{-\omega_2^2 e^{\tau_n(\omega_1+\omega_2)}\varphi_n''(z)}{(1+i)\tau_n^{12}}
$$
$$
=
\begin{vmatrix}
-[1]_1 & i[1]_1 & -ie^{\tau_n\omega_2}[1]_1 & e^{\tau_n\omega_1}[1]_1 \\
-ie^{\tau_n\omega_1 z}[1]_1 & ie^{\tau_n\omega_2 z}[1]_1 & ie^{\tau_n\omega_2(1-z)}[1]_1 & -ie^{\tau_n\omega_1(1-z)}[1]_1 \\
e^{\tau_n\omega_1}[1]_1 & e^{\tau_n\omega_2}[1]_1 & [1]_1 & [1]_1 \\
-e^{\tau_n\omega_1}[1]_1 & ie^{\tau_n\omega_2}[1]_1 & -i[1]_1 & -[1]_1
\end{vmatrix}
$$

$$
= \Big\{ ie^{\tau_n\omega_2 z} - e^{\tau_n\omega_2(1-z)}e^{\tau_n\omega_2 z} + (i-1)e^{\tau_n\omega_1(1-z)}e^{\tau_n\omega_2 z}\Big\} + O\left(n^{-1}\right).
$$

Moreover,

$$
\frac{-\omega_2^2 e^{\tau_n(\omega_1+\omega_2)}\varphi_n'(z)}{(i+1)\tau_n^{11}} = \frac{1}{\sqrt{2}}\Big\{ (i+1)e^{\tau_n\omega_2 z} + (1-i)e^{\tau_n\omega_2(1-z)}e^{\tau_n\omega_2}
$$
$$
- 2e^{\tau_n\omega_2}e^{\tau_n\omega_1(1-z)}\Big\} + O\left(n^{-1}\right).
$$

We note from (3.31) that

$$
\begin{cases}
e^{\tau_n \omega_2} = e^{i(n-1/4)\pi} + O\left(n^{-1}\right) = \dfrac{(1-i)(-1)^n}{\sqrt{2}} + O\left(n^{-1}\right), \\[2mm]
e^{\tau_n \omega_2 z} = e^{i(n-1/4)\pi z} + O\left(n^{-1}\right), \\[2mm]
e^{\tau_n \omega_1 z} = e^{-(n-1/4)\pi z} + O\left(n^{-1}\right).
\end{cases}
\tag{3.44}
$$

By setting

$$
f_n(x) = \frac{-\omega_2^2 e^{-1/4 \int_0^z a(s)ds} e^{\tau_n(\omega_1+\omega_2)} \varphi_n(z)}{(1+i)\tau_n^{12}},
\tag{3.45}
$$

the expression (3.40) can then be concluded. Furthermore,

$$
\frac{-\omega_2^2 e^{\tau_n(\omega_1+\omega_2)} \varphi_n'(z)}{(i+1)\tau_n^{12}} = O\left(n^{-1}\right),
$$

then

$$
f_n''(x) = \frac{\left(\dfrac{\rho(x)}{EI(x)}\right)^{1/2} \omega_2^2 e^{-1/4 \int_0^z a(s)ds} e^{\tau_n(\omega_1+\omega_2)} \varphi_n''(z)}{p^2(1+i)\tau_n^{12}},
\tag{3.46}
$$

then the expression (3.41) is obtained. Also, from (3.40), we have

$$
\zeta_n = m\lambda_n f_n(1) = me^{-1/4 \int_0^1 a(s)ds} \left[\sqrt{2}\cos(n-1/4)\pi - (-1)^n + O\left(n^{-1}\right)\right]
$$

$$
= O\left(n^{-1}\right),
\tag{3.47}
$$

also, we have from the boundary condition of (3.14), $\lambda_n J f_n'(1) = -\dfrac{EI(1)f_n''(1)}{\lambda_n J} -$
$cf_n'(1)$, we obtain $\delta_n = J\lambda_n f_n'(1) = O\left(n^{-2}\right)$. The theorem is proved. $\qquad\square$

3.4 Riesz Basis Property

Definition 3.1 Let $\mathbb{A}$ be a closed-loop operator in a Hilbert space $\mathbb{H}$. A nonzero element $x \neq 0 \in \mathbb{H}$ is called a generalized eigenvector of $\mathbb{A}$ corresponding to an eigenvalue λ (with finite algebraic multiplicity) of $\mathbb{A}$ if there exists a nonnegative integer n such that $(\lambda - \mathbb{A})^n x = 0$.

Definition 3.2 A sequence $(x_n)_{n\geq 1}$ in $\mathbb{H}$ is called a Riesz basis for $\mathbb{H}$ if there exists an orthonormal basis $(z_n)_{n\geq 1}$ in $\mathbb{H}$ and a linear bounded invertible $T \in \mathcal{L}(\mathbb{H})$ such that $Tx_n = z_n$ for any $n \in \mathbb{N}^*$.

Theorem 3.3 (See [2]) *Let $(\lambda_n)_{n\geq 1} \subset \sigma(\mathbb{A})$ be the spectrum of $\mathbb{A}$. Assume that each λ_n has a finite algebraic multiplicity m_n and $m_n = 1$ as $n > N$ for some integer N, then there is a sequence of linearly independent generalized eigenvectors $\{x_n\}_1^{m_n}$ corresponding to λ_n. If $\left\{\{x_n\}_1^{m_n}\right\}_{n\geq 1}$ forms a Riesz basis for $\mathbb{H}$, then $\mathbb{A}$ generates a C_0 semigroup $e^{\mathbb{A}t}$ which can be represented as*

$$e^{\mathbb{A}t}x = \sum_{n=1}^{+\infty} e^{\lambda_n t} \sum_{i=1}^{m_n} a_{ni} \sum_{j=1}^{m_n} f_{nj}(t)x_{nj},$$

for any $x = \sum_{n=1}^{+\infty} \sum_{i=1}^{m_n} a_{ni}x_{ni} \in \mathbb{H}$ where $f_{nj}(t)$ is a polynomial of t with order less than m_n. In particular, if $a^* < Re\lambda < b^*$ for some real numbers a^* and b^*, then $\mathbb{A}$ generates a C_0 group on $\mathbb{H}$. Moreover, the spectrum-determined growth condition holds $e^{\mathbb{A}t}: \omega(\mathbb{A}) = S(\mathbb{A})$, where

$$\omega(\mathbb{A}) = \lim_{t \to +\infty} \frac{1}{t} \| e^{\mathbb{A}t} \| \text{ is the growth order of } e^{\mathbb{A}t} \text{ and } S(\mathbb{A}) = \sup\{Re\lambda / \lambda \in \sigma(\mathbb{A})\}$$

is the spectral bound of $\mathbb{A}$.

In order to remove the requirement of the estimation of the low eigenpairs of the system, a corollary of Bari's theorem is recently reported in [2], which provides a much less demanding approach in generating a Riesz basis for general discrete operators in the Hilbert spaces. The result is cited here.

Theorem 3.4 (See [2]) *Let $\mathbb{A}$ be a densely defined discrete operator, that is, $(\lambda - \mathbb{A})^{-1}$ is compact for some λ in a Hilbert space $\mathbb{H}$. Let $\{z_n\}_1^{+\infty}$ be a Riesz basis for $\mathbb{H}$. If there are an $N \geq 0$ and a sequence of a generalized eigenvectors $\{x_n\}_{N+1}^{+\infty}$ of $\mathbb{A}$ such that*

$$\sum_{n=N+1}^{+\infty} \|x_n - z_n\|^2 < +\infty,$$

then

1. *There are an $M > N$ and generalized eigenvectors $\{x_{n_0}\}_1^M \cup \{x_n\}_{M+1}^{+\infty}$ form a Riesz basis for $\mathbb{H}$.*
2. *Consequently, let $\{x_{n_0}\}_1^M \cup \{x_n\}_{M+1}^{+\infty}$ correspond to eigenvalues $\{\sigma_n\}_1^{+\infty}$ of $\mathbb{A}$, then $\sigma(\mathbb{A}) = \{\sigma_n\}_1^{+\infty}$ where σ_n is counted according to its algebraic multiplicity.*
3. *If there is an $M_0 > 0$ such that $\sigma_n \neq \sigma_m$ for all $m, n \geq M_0$, then there is an $N_0 > M_0$ such that all $\sigma_n, n > N_0$ are algebraically simple.*

In order to apply Theorem 3.4 to the operator $\mathbb{A}$ when we consider $\{x_n\}$ in Theorem 3.4 as the eigenfunctions of $\mathbb{A}$, we need a referring Riesz basis $\{z_n\}_1^{+\infty}$ as well. For the system (3.1), this is accomplished by collecting (approximately) normalized eigenfunctions of the following free conservative system:

$$\begin{cases} \rho(x)y_{tt}(x,t) + (EI(.)y_{xx})_{xx}(x,t) = 0, & 0 < x < 1, \ t > 0, \\ y_x(0,t) = y(0,t) + (EI(.)y_{xx})_x(0,t) = 0, & t > 0, \\ (my_{tt} - (EI(.)y_{xx})_x)(1,t) = 0, & t > 0, \\ (Jy_{xtt} + EI(.)y_{xx})(1,t) = 0, & t > 0, \\ y(x,0) = y_0(x), \ y_t(x,0) = y_1(x), & 0 < x < 1. \end{cases} \tag{3.48}$$

The system operator $\mathbb{A}_0$ associated with (3.48) is nothing but the operator $\mathbb{A}$ with $b = c = \dfrac{1}{a} = 0$.

$$\begin{cases} \mathbb{A}_0(f, g, \zeta, \delta) = (g, -\dfrac{1}{\rho(.)}(EI(.)f'')'', \ (EI(.)f'')'(1), \ -EI(1)f''(1)), \\ D(\mathbb{A}_0) = \{(f, g, \zeta, \delta) \in (H^4(0,1) \cap \mathbb{V}) \times \mathbb{V} \times \mathbb{C}^2 / f(0) + (EI(.)y_{xx})_x(0) = 0, \\ \zeta = Jg'(1), \ \delta = mg(1)\}. \end{cases}$$
$$\tag{3.49}$$

$\mathbb{A}_0$ is skew-adjoint with compact resolvent in $\mathbb{H}$. It is seen that all the analyses in the previous sections for the operator $\mathbb{A}$ are still true for the operator $\mathbb{A}_0$. Therefore, we have the following counterpart of Theorem 3.2 for the operator $\mathbb{A}_0$:

Lemma 3.4 *Each eigenvalue υ_{n_0} of $\mathbb{A}_0$ with sufficiently large module is geometrically simple hence algebraically simple.*

The eigenfunctions $\overrightarrow{\Psi_{n_0}} = (f_{n_0}, \upsilon_{n_0} f_{n_0}, m\upsilon_{n_0} f_{n_0}(1), J\upsilon_{n_0} f'_{n_0}(1))$ *of* υ_{n_0} *have the following asymptotic expressions:*

$$\upsilon_{n_0} f_{n_0}(x) = e^{-1/4 \int_0^z a(s)ds}\left[\sqrt{2}\cos(n-1/4)\pi z - (-1)^n e^{-(n-1/4)\pi(1-z)} + O\left(n^{-1}\right)\right],$$
$$\tag{3.50}$$

$$f''_{n_0}(x) = \dfrac{1}{p^2}\left(\dfrac{\rho(x)}{EI(x)}\right)^{1/2} e^{-1/4\int_0^z a(s)ds}\left[\sqrt{2}i\cos(n-1/4)\pi z + i(-1)^n \right.$$
$$\left. \times e^{-(n-1/4)\pi(1-z)} + O\left(n^{-1}\right)\right],$$
$$\tag{3.51}$$

$$\zeta_{n_0} = O\left(n^{-1}\right), \ \delta_{n_0} = O\left(n^{-2}\right), \tag{3.52}$$

where all $(\upsilon_{n_0}, \overline{\upsilon_{n_0}})$, but possibly a finite number of other eigenvalues, are composed of all the eigenvalues of $\mathbb{A}_0$.

The eigenfunctions $\overrightarrow{\Psi_{n_0}} = (f_{n_0}, \upsilon_{n_0} f_{n_0}, m \upsilon_{n_0} f_{n_0}(1), J \upsilon_{n_0} f'_{n_0}(1))$ are normalized approximately.

From a well-known result in functional analysis, we know that the eigenfunctions of $\mathbb{A}_0$ form an orthogonal basis for $\mathbb{H}$, particularly, all $\overrightarrow{\Psi_{n_0}}$ and their conjugates form an (orthogonal) Riesz basis for $\mathbb{H}$.

Then there exists a positive integer large enough N such that

$$\sum_{n=N+1}^{+\infty} \left\| \overrightarrow{\Phi}_n - \overrightarrow{\Psi}_{n_0} \right\|_{\mathbb{H}}^2 = \sum_{n=N+1}^{+\infty} O(n^{-2}) < +\infty. \tag{3.53}$$

The same result is verified for their conjugates. We can now apply Theorem 3.4 to obtain the main results of the present paper.

Theorem 3.5 *Let the operator be $\mathbb{A}$ defined by (3.7).*

1. *There is a sequence of generalized functions properly normalized of $\mathbb{A}$ which forms a Riesz basis of the Hilbert space $\mathbb{H}$.*
2. *The eigenvalues of $\mathbb{A}$ have the asymptotic behavior (3.31).*
3. *All $\lambda \in \sigma(\mathbb{A})$ with sufficiently large modulus are algebraically simple. Therefore, $\mathbb{A}$ generates a C_0 semigroup on $\mathbb{H}$. Moreover, for the semigroup $e^{\mathbb{A}t}$ generated by $\mathbb{A}$, the spectrum-determined growth condition holds.*

As a consequence of Theorem 3.5, we have a stability result for system (3.1).

Corollary 3.1 *The system (3.1) is exponentially stable for any $a > 0$, $b \geq 0$, and $c > 0$.*

Proof Theorem 3.5 ensures the spectrum-determined growth condition: $\omega(\mathbb{A}) = sup\{Re\lambda : \lambda \in \sigma(\mathbb{A})\}$, Lemma 3.2 (in the proof of Corollary 3.1), say that $Re\lambda < 0$ provided $\lambda \in \sigma(\mathbb{A})$ and Theorem 3.1 shows that imaginary axis is not an asymptote of $\sigma(\mathbb{A})$. Therefore $sup\{Re\lambda : \lambda \in \sigma(\mathbb{A})\} < 0$. □

References

1. G. Chen, S. G. Krantz, D. W. Ma, C. E. Wayne, H. H. West, The Euler-Bernoulli beam equation with boundary energy dissipation *in Operator Methods for Optimal Control Problems*, 2nd edn. (S. J. Lee, ed., Lecture Notes in Pure and Appl. Math., Marcel Dekker, New York), **108** pp. 67–96
2. B. Z. Guo, R. Yu, On Riesz basis property of discrete operators with application to an Euler-Bernoulli beam equation with boundary linear feedback control. IMA J. Math. Control Inform. **18** pp. 241–251 (2001)
3. B. Z. Guo, R. Yu, Riesz basis property and exponential stability of controlled Euler-Bernoulli beam equations with variable coefficients. SIAM J. Control Optim. **40** pp. 1905–1923 (2002)
4. S. W. R. Lee, H. L. Li, Development and characterization of a rotary motor driven by anisotropic piezoelectric composite. SIAM J. Control Optim. Smart Materials Structures **7** pp. 327–336 (1998)

5. W. Littman, L. Markus, Exact boundary controllability of a hybrid system of elasticity. Arch. Rational Mech. Anal. Structures **103** pp. 193–236 (1988)
6. V. Komornik, Exact controllability and stabilization (The Multiplier Method) *Masson, Paris:Wiley* (1995)
7. M. A. Naimark, Linear Differential Operators, Part 1: Elementary Theory of Linear Differential Operators *Ungar Publishing Co., New York* (1967)
8. A. Pazy, Semigroups of linear operators and applications to partial differential equations *Springer-Verlag, New York* (1983)
9. B. Rao, Recent results in non-uniform and uniform stabilization of the Scole model by boundary feedbacks *Lecture Notes in Pure and Applied Mathematics, J.-E Zolesio, ed., Marcel Dekker, New York* **163**(1983) pp. 357–365 (1994)
10. J. M. Wang, G. Q. Xu, S. P. Yung, Riesz basis property, exponential stability of variable coefficient Euler Bernoulli beams with indefinite damping. IMA J. Appl. Math. **163** pp. 459–477 (2005)

Part II
Advanced Mathematics for Imaging

In this second part, we collect three chapters dealing with some applications of advanced mathematics to the broad area of signal processing and imaging.

The first chapter written by Serge Dos Santos begins with a review of some relevant mathematical tools such as Lie groups, correlation functions, and Fourier analysis, with focus on applications to Non-Destructive Testing (NDT) and harmonic medical imaging. The second chapter written by Thomas Deregnaucourt, Chafik Samir, Abdelmoujib Elkhoumri, Jalal Laassiri, and Youssef Fakhri describes a geometrically constrained manifold embedding for an extrinsic Gaussian process applied in the medical context where Ultrasound (US) and Magnetic Resonance Imaging (MRI) techniques are noninvasive testing (NIT). The third chapter written by Houda Salmi, Khalid El Had, Hanan El Bhilat, and Abdelilah Hachim gives a numerical analysis of the stress intensity factor for a crack in P265GH steel.

Chapter 4
Advanced Ground Truth Multimodal Imaging Using Time Reversal (TR) Based Nonlinear Elastic Wave Spectroscopy (NEWS): Medical Imaging Trends Versus Non-destructive Testing Applications

Serge Dos Santos

Abstract An innovative pragmatic approach is described using a symbiosis of modern signal processing techniques coming from nonlinear science and multiscale global analysis usually considered for the characterization of nanoscale systems with complex mesoscopic properties. The basis of this systemic approach comes from ultrasound imaging of the complexity with examples coming from the non-destructive testing (NDT) industry and from the medical harmonic imaging research. Modeling uses basic tools derived from the analysis of nonlinear dynamics, such as spectrum representations coming from harmonic analysis and multiscale analysis, in association with advanced signal processing such as similarity and invariance analysis based on group theory. Experimentation is based on multimodal imaging using time reversal (TR) based nonlinear elastic wave spectroscopy (NEWS). Practical implementation in the aeronautic industry and the biomedical imaging field is taken as a demonstration of feasibility of such advanced engineering TR-NEWS methods for imaging complex systems with an advanced ground truth approach including round robin tests. This new class of multiscale materials is studied with phenomenological approaches like the Preisach–Mayergoyz space (PM space), where some physical properties like hysteresis, end-point memory, and odd harmonic generations are extracted from the noise. As a consequence, this approach is proposed for including phenomenological tools to any organizations' strategies in order to monitor the vulnerability in a societal emergency management context. These results could be applied for the organizational side of innovation, cognition,

S. Dos Santos (✉)
Institut National des Sciences Appliquées, Centre Val de Loire, Blois, France

U1253 "Imaging and Brain: iBrain", Team Imaging, Biomarkers, Therapy, Inserm, Université de Tours, Tours, France
e-mail: serge.dossantos@insa-cvl.fr

and practices for novelty and disruption in political organizations, social monitoring, and any system which will involve big data or artificial intelligence.

Keywords Signal processing · Symmetry · Invariance · Nondestructive testing (NDT) · Ultrasonic imaging · Multiscale · Pragmatic probability law · Integritology · Hierarchical analysis · Decision theory · Ground truth · TR-NEWS

4.1   Introduction to Non-invasive and Non-destructive (NDT) Testing

Monitoring multiscale systems with multimodal imaging is now recognized to be a keystone for understanding aging and degradation processes in complex modern materials and biological media. After more than 90 years of non-destructive testing (NDT) applications, significant progress has been made from the research laboratory to the manufacturing floor [1–5]. Today, there are numerous excellent NDT capabilities available to the user. In many cases, these capabilities are not necessarily being utilized appropriately due to lack in education and training or in the culture of NDT. NDT is and will continue to be a key science in addressing the complex needs for reliability and safety in the construction and service life of structures. Consequently, it directly impacts people and their daily life. Design engineers who develop projects need a priori knowledge of NDT technology. They need to determine the appropriate NDT method to be performed during construction and incorporate them into their designs. During their service life, most constructions are subject to different types of degradation of their structural integrity, such as corrosion, fatigue, aging, etc. In many cases, the consequences have been catastrophic resulting in train derailments, mid-air aircraft failures, bridge collapses, pipelines and refinery explosion, and offshore platform petroleum spills. NDT integrity engineering is a discipline to develop non-destructive testing and evaluation involving materials science, fracture mechanics, and other sciences that would guarantee and enhance the reliability and safety by ensuring integrity of structures in everyday life.

The understanding of the complexity of the aging of biological media is still a challenge. Improved imaging and non-invasive testing (NIT) devices need to be continuously proposed in order to extract physical information from applications of image processing that gives the best knowledge of the complex system. Image thresholding techniques are used in both medical imaging area and NDT [6–8]. In these contexts, advanced ground truth should be applied. Ground truth allows image data to be related to real features and materials on the ground (Fig. 4.1). The collection of ground-truth data enables calibration of remote-sensing data, and aids in the interpretation and analysis of what is being measured. Other applications include cartography, meteorology, analysis of aerial photographs, satellite imagery, and other techniques in which complex data are gathered at a distance. In the case of a classified image, it allows supervised classification [9] to determine the accuracy of the classification performed using various algorithms.

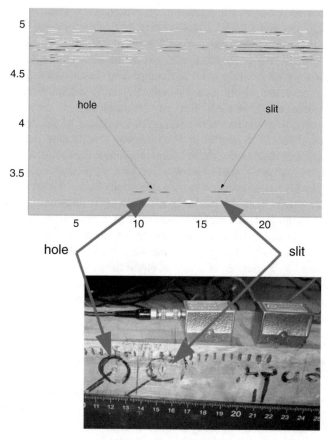

Fig. 4.1 Ground truth images applied in NDT with an example of round robin tests. As ground truth images for the training, validation, and evaluation of slit and hole, complex imaging methods should be used in symbiosis with signal and image processing

In this engineering area devoted to complexity, the high number of parameters (and their associated precision and uncertainty) induces an increase of the stochastic part of the physical information, decreasing consequently the deterministic part [10].

Accurate analysis of complex systems needs the use of new and powerful methods of signal processing [11], dealing with stochastic data. Another property that is commonly related to complex systems is that it shows nonlinearities, which is closely related to the concept of chaos. In biological systems with nonlinear signatures, small causes might have large effects. Such systems might be very sensitive to the initial conditions and/or very sensitive to the excitation properties, i.e., stability, vulnerability, or accuracy, for example. Consequently, a small initial difference in excitation might lead to large differences in the subsequent response of the system. This is frequently referred to as deterministic chaos. A given initial system state might lead to several different final states, impossible in principle to know in advance which of the states the system ends up. Most of systems used in engineering presents a level of nonlinearity that was considered negligible and included in the small stochastic part of the noise. Now, modern engineering is developed by considering this stochastic part of the nonlinear signature as a new vector of information coming from the complex system under study [12]. With the current innovations in big data sciences and artificial intelligence (AI), these signatures should be used as a new family of information.

Consequently, since the structure of complex biological, social, and professional organizations is known to exhibit memory effects, conditioning, local and global synchronization, hysteresis, threshold effects, amplitude dependence, and saturation at various states, an extension of a mesoscopic signal processing could be introduced in order to quantify the new proportion of stochasticity in the (nonlinear) response of any complex system. Since stochastic signals are actually produced by deterministic mesoscopic systems that are capable of nonlinear stochastic responses, their behavior should be associated with invariant properties, such as symmetries like time invariance and stationarity. For example, the stability of such mesoscopic system is also conditioned by a complex skeleton of elementary rules or elements which are necessary for synchronized behavior. Under external excitations, nonlinear systems (or complex organizations) can produce nondeterministic responses which increase the stochastic part with a non-intuitive proportion that needs to be considered in biology, modern engineering organizations, and social monitoring, in order to understand the breaking of synchronized states.

The huge variety of information extracted from this small stochastic part of the response coming from a complex system induces an increase of the uncertainty associated with the linear part. This linear part, with its underlying hypothesis of stationarity and determinism, should be consequently associated with a greater uncertainty if the system under study presents intrinsically a complex structure with mesoscopic properties, memory effects, conditioning, and aging processes [13, 14], including the conservation and rehabilitation of pieces of the historical and cultural heritage like stones, paintings, frescos, mosaics, jewelry, or any objects fabricated by our ancient civilization [15]. Of course, these properties are breaking now the stationarity hypothesis implicitly assumed in any linear signal processing, since linear systems theory dominates the field of engineering.

Since the last two decades, the NDT community developed, at the research level first but now at the industrial level, a new class of signal processing tools for extracting the nonlinear signature of damaged materials [16, 17]. The nonlinear elastic waves spectroscopy (NEWS) methods were developed with signal processing improved for extracting, from the complex materials and systems, new nonlinear deterministic signatures [18, 19]. These signatures resulting from a nonlinear mixing of waves (Fig. 4.2) allow information about the nonlinearity of medium: harmonics and modulation for weak (classical) nonlinearity [20–22]; slow dynamics, subharmonics, hysteresis and memory effects for strong (non-classical) nonlinearity observed in mesoscopic materials [23, 24].

Recently, the use of methods, approaches, and results coming from the huge domain of nonlinear physics has increased the number of industrial applications in the engineering industry. In order to illustrate this approach, this fruitful domain

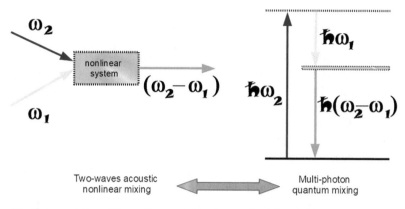

Fig. 4.2 Representation of the nonlinear mixing of waves (optic/acoustic)

of NDT and particularly the improvement of modern methods using nonlinear ultrasound has be taken as an example of a complex system which needs systemic approaches that could be transposed for defining new emerging entrepreneur. Systemic analysis is a systems-based framework devoted to the structured analysis of complex systems understanding. With a systemic view, complex systems can be described with succinct parameters which are intrinsically linked to its generalized symmetries (Fig. 4.2). For example, fractal and multifractal concepts, usually used in the analysis of nonlinear systems [25] provide a systemic description of the complex dynamic of nonlinear waves.

During the last 10 years, the principle of an a priori use of symmetries and similarity properties seems to have an increase interest for signal processing applied to nonlinear acoustics phenomena [26]. Furthermore, several models and equations have been analyzed with theoretical methods, intrinsically based on Lie groups properties, like in quantum physics and nonlinear optics as represented in Fig. 4.2. For example, new signal processing methods have been validated in NDT and for harmonic imaging using an ultrasound contrast agent (UCA) [27]. Among them, pulse inversion (PI) techniques have been extended and generalized using symmetry analysis [28].

Invariance with respect to time is one of the properties of a more general algebraic approach that is applied in physics which uses intrinsic symmetries for the simplification and the analysis of complex systems. Symmetry analysis [29] is the framework of a systemic approach aimed at using absolute symmetries like time reversal (TR), reciprocity between emitters and receivers, and others. The idea of including advantages of TR and reciprocity invariance in NEWS was motivated by experimental results obtained on bubbly liquids [30, 31]. In recent years there has been a considerable development of TR based NEWS methods using invariance with respect to TR and reciprocity, both in numerical and experimental settings [1]. As a fruitful example, TR-NEWS systemic methods have the potential to become a powerful and promising tool for the NDT industry. They provide the objective to detect and image structural damages in complex medium, thanks to the use of advanced signal processing techniques based on multiscale analysis applied to a big amount of data. Such nonlinear TR methods are now recognized in other domains such as cryptography [32] and ultrasonography for dentistry [33] for localization or the elastic nonlinearity in complex medium [34]. These concepts were practically elaborated as the TR-NEWS methods [35, 36]. Using symbiosis of these systems, TR-NEWS fundamental experimental demonstrations [35] have been conducted with applications in the improvement of nonlinear scatterers identification, such as bubbles [30], landmines [37], cracks in complex aeronautic materials [36] and is now highly recognized as extremely reliable [38, p. 14]. New excitations are now in progress in order to give to TR-NEWS methods the practicability needed for both the NDT and the medical imaging community [39] where the term NIT is preferred.

Finally, symmetry preservation is one of the most important properties that a turbulence model should possess, since the symmetry group translates mathematically the physical properties of the flow. In fluid mechanics, the symmetry theory helps to exhibit similarity properties and to deduce scaling laws, such as algebraic, logarithmic (wall), or exponential laws. For example, Burgers and Earnshaw equations, which describe one- dimensional propagation of acoustic waves, have been studied with symmetry analysis and well-known transformation (such as Hopf–Cole for Burgers). The study can be revisited with the symmetry analysis point of view. In this paper, the methodology of symmetry analysis is presented. Some simple demonstration of calculation conducted on nonlinear acoustics equation such as Burgers and non-classical model describing the acoustic propagation of a pulse in a medium with nonclassical nonlinearity confirms the interest of this method in signal processing. For example, it was shown [40] that the symmetry properties of a general Lienard type equation exhibit TR symmetry within the contributions of Lie infinitesimals. The discrete Lie symmetries related to TR will be presented from the theoretical and the experimental point of view.

The objective of this paper is to establish or maintain a fruitful network among scientists coming from basic sciences of complex system engineering (Technology, Engineering, and Mathematics), but also with researchers coming from social sciences and natural sciences where NDT and NIT are the basic stones of the integrity discipline, i.e., the integritology. Thanks to these collaborations, a good understanding of complex systems will lead to an increase of information, leading to new methods, new techniques, and new equipment with their multimodality property. Examples will concern the research conducted in the field of nonlinear acoustics for NDT applied to damaged materials of the aeronautic industry, and convey the potential of advanced signal processing for understanding complex systems. Some new results concerning the ultrasound based NEWS multimodal imaging and its improvement will be considered as an example of multimodality and complexification of instrumentation for nonlinear imaging of biological media with the non-invasive requirement.

The systemic analysis proposed in this paper can be seen as a world-view where standard linear approaches are supplemented by the framework of multiscale and nonlinear analysis. The final motivation of these results is to promote complex system engineering encouraging the application of the findings in the field of NDT integrity engineering [41]; in engineering departments, medical centers and institutions, and other relevant structures throughout the world. Finally, an extension is proposed for applying pragmatic approaches for organization's strategies and the study of social behavior. This non-classical analysis consists in assuming a high level of stochastic hierarchy in real situations: these real cases being frequently observed in complex organizations, climatology, and social sciences.

4.2 Memory Effects, Aging, and Nonlinear Viscoelastic Multiscale Behavior

In soft matter physics, environmental mechanics, non-Newton fluid mechanics, viscous elastic mechanics, porous media dynamics, and anomalous diffusion, fractional order differential research have attracted lots of interest like wavelet methods [42–44]. Moreover, it has been reported that the diseased and pathological tissues have larger nonlinear response than corresponding normal tissues, and the methods of nonlinearity measurements were further extensively investigated in many works, including fractional modeling of multiscale viscoelasticity [42].

One of the most difficult properties in biomaterial to measure, evaluate, and model is the multiscale viscoelastic effect. Several authors tried to extract some local properties using specific experimental setup: uniaxial loadings, biaxial protocols, 1D, 2D, 3D modeling. Furthermore, several effects such as Mulling effects or relaxation were identified in all the communities and seem to be connected all together. For example, in [45], the viscous property of the skin is shown to be influenced by the stress relaxation process. In our approach, all these viscoelastic effects could be associated with multiscale pragmatic and phenomenologic parameters that could be modeled by fractional models where viscoelasticity appears naturally as a direct consequence of the multiscale property of the skin [46, 47]. A basic consequence of viscoelasticity is the energy losses which are difficult to evaluate in biological medium. Not only the absolute amplitude of energy is difficult to measure, but also the nature of this energy: mechanical, thermal, chemical, etc. Since the multiscale properties are assumed in our approach (Fig. 4.3), energy transfer flux should also be present at all scales, and should induce physical phenomenon at all scales, from the mechanical domain (at low frequency) to the acoustical domain (at 20 MHz, involving solid–solid, solid–fluid, and fluid–fluid interaction at the mesoscopic scale). Energy losses appear also in all medium showing hysteresis effects [48, 49]. Losses are usually associated with the hysteresis area describing the excitation-response curve of the material (Fig. 4.4). The analogy between the memory effects of the memristor [50, 51] and hysteresis effects in the skin allows us to suggest the same physical origin of aging [52–55]. In [56, 57], we consider memristors as a plausible solution for the realization of transducers as an autonomous linear time variant system for TR-NEWS applications, especially for measuring non-classical nonlinearities.

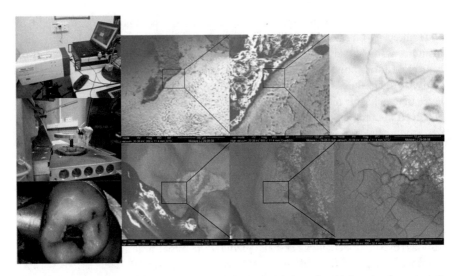

Fig. 4.3 Multimodality in the imaging of multiscale properties of a third molar human tooth surface with a Polytec laser vibrometer, scanning electron microscope (ESM), and Hirox 3D digital microscope (left) showing the complex distribution of the cracks which connect the 1–5 μm cylindrical tubules. Size, distribution, and statistical properties of cracks of the healthy tooth (top) are compared with those of the damaged tooth (bottom)

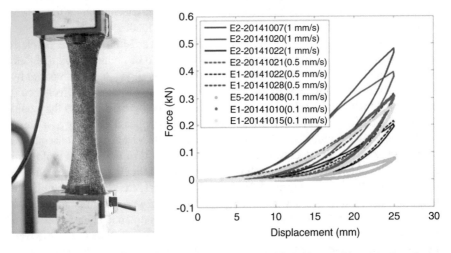

Fig. 4.4 Force–displacement curve characteristics of the studied skin samples versus the velocity of the loadings ($V = 0.1$; 0.5 and 1 mm/s). All samples are $100 \times 30 \times 2$ mm size

Such highly nonlinear behavior produces strongly nonlinear frequency spectrum broadening inducing low frequency effects equivalent to long time like reverberation and slow dynamics. This consequence induces naturally the problem of duration of any experiment showing this phenomenon. This phenomenon could explain the high number of relaxation parameters present in any uniaxial loading [58]. In our approach, by including memory effects (or memristive effects) in the multiscale phenomenological model, it is assumed that long time behavior could be naturally evaluated by assuming a statistical distribution of parameters [59]. Consequently, all slow dynamic phenomena are considered even if their measurements cannot be practically conducted in experiments. This multiscale approach focuses on the creation of the intelligent agent based reporting multiscale experiments that produce the ground truth necessary for data fusion to analyze the multiscale memristive properties of any aged medium [60, 61]. This reporting experimental process is a crucial first step in the construction of a wide big data system to produce a data fusion support tool for evaluating the aging of any system.

4.3   Multiscale Analysis and Hierarchical Structure

In this part, we will present an example of a multiscale NEWS analysis with signal processing methods for NDT applications. Complex properties of damaged materials can be presented with the Preisach–Mayergoyz space (PM space) phe-nomenological approach [16]. The consequences of the PM space modeling of mesoscopic materials will be highlighted with several examples [59, 62]: ultra-sonic measurements in damaged materials, electronic characterizations of complex network of hysteresis relays, and skin identification (Fig. 4.5). This phenomeno-logically based approach is under investigation for the conception of new kinds of system identification, validated and patented for complex biological systems such as bone, tooth [63], or complex manufactured products coming from the bio and agronomy industries. The objective to extend this modern approach to skin [58, 64]

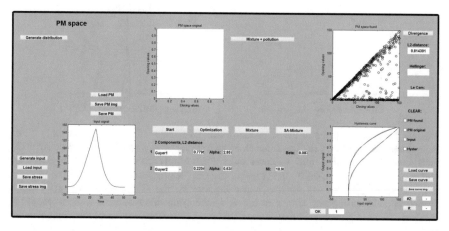

Fig. 4.5 Identification of the PM space along the mixture of Guyer 1 and Guyer 2 distribution, value of the L_2-distance is equal to 0.814

and human brain, from whose memory effects are currently admitted, gives to this big data analysis a promising future for modern engineering and medical imaging.

The consequence is a pragmatic analysis swarming by phenomenological approaches in the family of PM models. The accurate extracted information coming from such systems needs to be associated with the symmetry of the underlying mesoscopic structure with respect to scaling effects responses. In order to extract nonlinear signatures of such hierarchical structure, complex loadings (chirp-coded excitations or pulse compression [65]) are needed for their analysis which is highly multiscale in the frequency domain. Advanced and optimized signal processing is consequently mandatory for an accurate extraction of the information [66]. Modern ultrasonic imaging [67] inducing multimodality is finally requested for end-users and also demands speed and accurate embedded digital signal and image processors. Multimodal imaging coming from the modern medical imaging is now a universal project inducing new approaches [68] and new imaging strategies (Fig. 4.6).

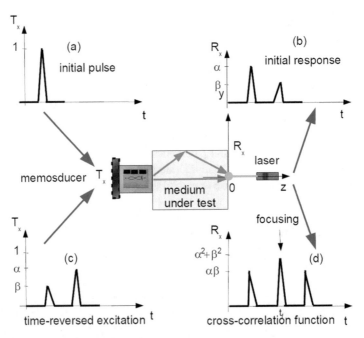

Fig. 4.6 Schematic process of the memristor based TR-NEWS with the virtual memory (time-delays) transducer concept. (**a**) The initial broadband excitation $T_x(t)$ propagates in a medium. (**b**) Additional echoes coming from interfaces and scatterers in its response R_x could be associated with a virtual source $T_x^{(2)}$. (**c**) Applying reciprocity and TR process to R_x. (**d**) The time reversed new excitation $T_x = R_x(-t)$ produces a new response R_x (the TR-NEWS coda $y_{TR}(t)$) with a spatiotemporal focusing at $z = 0$; $y = 0$; $t = t_f$ and symmetric side lobes with respect to the focusing. This is a physical interpretation of the cross-correlation function of the medium under test

4.3.1   TR-NEWS Signal Processing

The roles of the transducers are not changed during the experiment: the focusing of the ultrasonic wave relies on the TR-NEWS signal processing [69]. This is a two-pass method where the receiving and transmitting transducers do not change their roles. In this sense, the "Time Reversal" describes the signal processing method which accounts for internal reflections of the material as virtual transducers, used for focusing the wave in the second pass of the wave transmission. The placement of the transducers is not important from the signal processing standpoint: in NDT investigation they could be placed arbitrarily and they do not have to be in line with each other, but the configuration must remain fixed during the complete TR-NEWS procedure. Figure 4.6 outlines the TR-NEWS signal processing steps for an initial pulse excitation using the memosducer concept [56]. In order to increase the signal-to-noise ratio [70], a chirp excitation is sent instead of a pulse of figure (a), inducing the same "frequency content" in the signal. Then this chirp-coded excitation $c(t)$ is transmitted through the medium, with

$$c(t) = A \sin (\psi(t)), \tag{4.1}$$

where $\psi(t)$ is linearly changing instantaneous phase. Typically, a linear sweep from 0 to 10 MHz is used. Then the chirp-coded coda response $y(t)$ with a time duration T is recorded at the receiver

$$y(t, T) = h(t) * c(t) = \int_{\mathbb{R}} h(t - t', T)c(t')dt', \tag{4.2}$$

where $h(t - t', T)$ is the impulse response of the medium. The $y(t, T)$ is the direct response from the receiving transducer when the chirp excitation $c(t)$ is transmitted through medium. Next the correlation $\Gamma(t)$ between the received response $y(t, T)$ and chirp-coded excitation $c(t)$ is computed during some time period Δt, with

$$\Gamma(t) = \int_{\Delta t} y(t - t', T)c(t')dt' \simeq h(t) * c(t) * c(T - t, T), \tag{4.3}$$

where the $h(t)*c(t)*c(T-t, T)$ is the pseudo-impulse response. Assuming $\Gamma_c(t) = c(t) * c(T - t) = \delta(t - T)$, it is proportional to the impulse response $h(t)$ if using linear chirp excitation for $c(t)$. Therefore the actual correlation $\Gamma(t) \sim h(t)$ contains information about the wave propagation paths in complex media.

Time reversing the correlation $\Gamma(t)$ from the previous step results in $\Gamma(-t)$ used as a new input signal. Re-propagating $\Gamma(-t)$ in the same configuration and direction as the initial chirp yields

$$y_{TR}(t, T) = \Gamma(T - t) * h(t) \sim \delta(t - T), \tag{4.4}$$

where $y_{TR} \sim \delta(t - T)$ is now the focused signal under receiving transducer where the focusing takes place at time T. Due to the fact that $\Gamma(t)$ contains information about the internal reflections of the complex media, its time reversed version $\Gamma(T-t)$ will eliminate these reflection delays by the time signal reaches the receiver, resulting in the focused signal y_{TR} (Eq. (4.4)). The test configuration must remain constant during all of these steps [71]; otherwise, the focusing is lost. The steps of this focusing process in a physical experiment are shown in Fig. 4.6. In this case, pulse inversion (PI) is an established method for detecting nonlinearities [28]. The procedure used here involves conducting TR-NEWS measurements with positive and negative sign for A in Eq. (4.1) and comparing the focused signals. Differences could indicate the presence of nonlinearities.

In terms of signal processing, applying the PI method by changing $c(t)$ into $-c(t)$ preserves the response $y_{TR}(t, T)$ invariant. Combining both PI and TR processes, intrinsic nonlinear signatures can be extracted experimentally by the symmetry of discrete processes. All this theory is valid under linear behavior of the medium represented by its impulse response $h_{21}(t)$. Any source of nonlinearity in the system will result in a perturbation of this method, and will induce additional terms in Eqs. (4.3)–(4.4). When advanced signal processing methods have extracted the nonlinear signature, the next step consist in assuming it as coming from set of localized sources. Moreover, discrete symmetries, i.e., symmetries associated with a discrete finite group, are very important in quantum mechanics. In this field one speaks of parity, charge conjugation, rotations, and by TR and one uses discrete symmetries to provide selection rules. We will now discuss two methods of determining the discrete symmetries of differential equations. One of them was proposed by Hydon [72] and it is essentially the classical method of constructing the normalizer of a group. The other is a modification of Lie's method, defining a discrete symmetry as a discretization of the parameter of a continuous symmetry.

4.4   Lie Groups and Symmetries for Nonlinear Systems

Lie group theory (or Lie symmetry analysis) is used in many different areas of nonlinear sciences due to its practical applications and the insight that it brings to describe physical systems [73–75]. In this regard, one of the main advantages of symmetry analysis is that symmetry properties of linear or nonlinear equations can be exploited to achieve simplifications for finding solutions or properties. In the case of differential equations, these simplifications could be in the forms of order and/or dimension reduction. In terms of signal processing and system analysis, symmetries also represent fundamental information regarding conservation laws that describe a physical phenomenon. As a result, knowledge of symmetry properties of equations describing nonlinear and complex systems often opens alternative pathways to define optimized excitations for approaching problems when looking for solutions. Therefore, it is usually a good practice to analyze symmetry properties of equations before a solution strategy (whether it is analytic or numeric) is decided.

Hydon [72] provides a clear method for calculating all of the discrete symmetries of a given differential equation. Hydon's methodology works as follows: first, find the Lie algebra of point symmetry transformation generators for the differential equation in question. These symmetry generators will give all of the symmetries continuously connected to the identity component but will not directly give the discrete symmetries.

In order to suggest a generalization to nonlinear systems [76], let us consider examples from linear and nonlinear propagations/diffusion of acoustic/thermal fields in a simple medium. As a first example, let us consider normalized temperature $T(t, x)$ in a one-dimensional semi-infinite medium is given by the heat equation

$$\mathcal{E}(T, t, x) = T_t - T_{xx} = 0. \tag{4.5}$$

A transformation $\tilde{x} = F(x, a)$ with $x = F(x, a_0)$, with the group parameter a, its infinitesimals $\xi(x)$ have the property $F(x, a) = e^{(a-a_0)\xi(x)}$. For a partial derivative equation or a differential derivative equation at order $n > 1$, it is possible to find a set of infinitesimals using symmetry properties of equation $\mathcal{E}$ [29]. Infinitesimals ξ_i (resp. η_i) correspond to independent (resp. dependent) generators of $\mathcal{E}$ which verify $dx_1/\xi_1 = \ldots = dx_p/\xi_p = dy_1/\eta_1 = \ldots = dy_q/\eta_q$, where x_i are independent variables and y_i are dependent ones. Extraction of $n - 1$ invariants of equation $\mathcal{E}$ simplifies the equation by decreasing its order or its number of independent variables. Applying symmetry analysis [66] to the heat equation leads to infinitesimals given by

$$\xi_1(t, x, T) = C_1 + C_2 t + C_3 t^2, \tag{4.6}$$

$$\xi_2(t, x, T) = C_2 x/2 + C_3 tx + C_4 + C_5 t, \tag{4.7}$$

$$\eta(t, x, T) = C_3(-tT/2 - x^2 T/4) - C_5 x T/2 + C_6 T + g(t, x), \tag{4.8}$$

where C_i's are constants and $g(t, x)$ a solution of (4.5). The associated Lie generators are

$$X_1 = \partial_t \; ; \; X_2 = \partial_x, \tag{4.9}$$

$$X_3 = T \partial_T; X_5 = t \partial_x + x T/2 \partial_T, \tag{4.10}$$

$$X_6 = t^2 \partial_t + tx \partial_x - \left(t - x^2/2\right) T/2 \partial_T, \tag{4.11}$$

$$X_4 = t \partial_t + x/2 \partial_x. \tag{4.12}$$

With X_4 one obtain $\frac{dt}{t} = \frac{dx}{x/2}$, which is related to the invariant $\zeta = \frac{x}{\sqrt{t}}$. The partial derivative equation $\mathcal{E}(T, t, x)$ is reduced to an ordinary differential equation $v''(\zeta) + \zeta v'(\zeta) = 0$ with the new variable $T(t, x) = v(\zeta)$, where ζ is extracted with symmetry properties, and where $v(\zeta) = A \int_0^{\zeta/2} e^{-y^2} dy = A \int_0^{\frac{x}{2\sqrt{t}}} e^{-y^2} dy = T(t, x)$. The similarity variables are then given by $\bar{x} = \frac{x}{1-at}$, $\bar{t} = \frac{t}{1-at}$, and

$\overline{T} = T\sqrt{1-at}\exp\left(-\frac{ax^2}{4(1-at)}\right)$ which can be used, for generating the nontrivial solution $T = \frac{T_0}{\sqrt{1+at}}\exp\left(-\frac{ax^2}{4(1+at)}\right)$ from the trivial constant solution T_0 and by replacing a by $-a$ analogue to a TR process. It is also used to determine particular solutions, called invariant solutions, or generate new solutions, once a special solution is known, in the case of ordinary or partial differential equations. The name similarity variables is due to the fact that the scaling invariance, i.e., the invariance under similarity transformations, was one of the first examples where this procedure has been used systematically. For each ordinary or partial differential equation ($\mathcal{E}$), one can find intrinsic symmetries. For example, linear equations $\mathcal{E}(x) = 0$ like the Heat equation have a symmetry $F(x) = Cx$, thanks to superposition principle. The interest of symmetry analysis comes for its invariance with respect to group properties.

As a second example, let us consider Lie group properties of acoustic wave propagation in hysteretic media [66]. Applying the Moran method of Lie reduction [77] to the hysteretic equation:

$$\mathcal{H}(V, V_i, V_M, \theta, \xi) = (V_\xi)^2 + V_M V_\xi V_\theta + V(V_M - V)(V_\theta)^2 = 0 \text{ where } i \in \{\theta, \xi\}, \tag{4.13}$$

where V, V_i, V_M, θ, ξ are given in reduced variables. One begins to build a one-parameter (a) group transformation $\overline{S} = C^s(a)S + K^s(a)$, where the letter S relates to all variables θ, ξ, and functions V, V_M, and $C^s(a), K^s(a)$ are group functions only dependent on the group parameter a. The derivative transformations give $\overline{S_i} = C^s/C^i S_i$, and $\overline{S_{ij}} = C^s/C^i C^j S_{ij}$. Any equation $\mathcal{H}(S, S_i, S_{ij})$ of variables S and its derivatives $S_i, S_{ij}, \ldots$ is said to be invariant if $\mathcal{H}(\overline{S}, \overline{S_i}, \overline{S_{ij}}) = F[a]\mathcal{H}(S, S_i, S_{ij})$, where $F[a]$ is a function which depends only on the group parameter a. This transformation forms a local group of point transformations establishing a diffeomorphism on the space of independent and dependent variables, mapping solutions of the equation to other solutions. Any transformation of the independent and dependent variables in turn induces a transformation of the derivatives. If we suppose that the solution of hysteretic equation is given by $V(\xi, \theta) = V_M(\xi)g(\eta, \theta)K\theta$, with K constant, then and after tremendous calculus involving implicit expressions [66], the solution $V(\xi, \theta)$ satisfies hysteretic equation if

$$-K\alpha\theta + (\beta + \alpha\xi)V(\xi, \theta)\left(K + \alpha\left((\beta + \alpha\xi)^{-\frac{\gamma}{\alpha}}V(\xi, \theta)\right)^{\frac{\alpha}{\gamma}}\right) = 0. \tag{4.14}$$

This new equation is expressed as a polynomial expression. For $n = \frac{\alpha}{\gamma} = 1$, a trivial exact explicit solution is obtained as follows: $V(\xi, \theta) = \frac{-\alpha + K\alpha\theta}{K(\beta + \alpha\xi)} = \frac{\alpha\theta - \alpha/K}{\alpha\xi + \beta}$, and can be used to generate an infinite number of solutions by applying a "nonlinear superposition principle."

Finally, generalized experimental TR based NEWS methods and their associated discrete symmetries can be taken as generic examples. These methods, already adopted as a powerful tool for several applications (NDT and medicine), are efficient approaches for focusing acoustic energy or energy localization [78] in a complex propagation medium: noisy, scattering, and complex shape medium [79]. Nonlinear Time Reversal acoustics is now an efficient technique for the improvement of localization of nonlinear scatterer coming from structural defects [1, 80–82]. Improvement of TR-NEWS is conducted with coded excitation using chirp frequency excitation and the concept was presented and validated in the context of NDE imaging [12]. The chirp-coded TR-NEWS method uses TR for the focusing of the broadband acoustic chirp-coded excitation.

4.4.1 Symmetries for Nonlinear TR Discrete Processes

As a first example, let us consider the Burger's equation

$$\mathcal{B}(u, t, x) = u_t + u u_x - u_{xx} = 0, \tag{4.15}$$

which has a five-dimensional Lie algebra where the associated Lie generators are $X_1 = \partial_t$; $X_2 = \partial_x$, $X_3 = x\partial_x + 2t\partial_t - u\partial_u$; $X_4 = 2t\partial_x + 2\partial_u$, $X_5 = tx\partial_x + 4t^2\partial_t + 4(x - tu)\partial_u$. Again, it can be shown [72] that the inequivalent complex discrete symmetries of Burgers' equation form a group which is generated by

$$\Gamma_1 : (x, t, u) \longmapsto (-ix, -t, iu) \text{ and } \Gamma_2 : (x, t, u) \longmapsto (x/(2t), -1/(4t), 2(tu-x)), \tag{4.16}$$

where the TR discrete process generated by Γ_1 can be seen as a Lie group transformation.

As a second example of discrete symmetries, modeling of ultrasound contrast agents (UCA) was done in order to define optimized excitations for TR-NEWS experiments applied to harmonic imaging [26, 83]. We analyzed UCA like oscillators coupled to their nearest neighbors. They form a lattice or chain with step h, supporting waves. Consider that the lattice is subjected to a parametric forcing with amplitude A and frequency ω_e, its motion is described by

$$\mathcal{U}(u, t, x) = u_{tt} + (\omega_0^2 + \eta \cos \omega_e t) \sin u - k^2(u_+ - 2u + u_-) = 0, \tag{4.17}$$

where u is the position of the ith element, $u_\pm = u(x \pm h, t)$, $\eta = 4\omega_e^2 A/L$ is the forcing parameter, and k is a constant denoting the strength of the coupling. For example, it has been shown [84] that for the Toda equation $\mathcal{T}(u, t, x) = u_{tt} - \exp(u_+ - u) - \exp(u - u_-) = 0$, the associated Lie generators are

$$X_1 = \partial_u \; ; \; X_2 = \partial_x, \tag{4.18}$$

$$X_3 = \partial_t; X_4 = t\partial_u, \tag{4.19}$$

$$X_5 = t\partial_t - \frac{2x}{h}\partial_u. \tag{4.20}$$

By using X_5, it can be shown that the transformations $\bar{t} = \mu t + \alpha$, $\bar{x} = \mu\nu x + \beta$, and $\bar{u} = \mu\nu u + \gamma$ induce the only admissible solutions given by $\mu = \pm 1$, $\mu\nu = \pm 1$ inducing the TR invariance when $\mu = -1$, $\nu = +1$ and the discrete symmetry given by the transformation $\bar{t} = -t$, $\bar{x} = x$, and $\bar{u} = u$. Consequently, there is a need in considering discrete symmetries involving nonlinear time reversal processes.

4.4.2   Universality of Nonlinear Systems and the Associated Signal Processing

In many of classical nonlinear systems, the distinction between prediction and explanation is often blurred. It makes perfectly sense that some phenomena are impossible, in principle, to predict, e.g., due to inherent stochasticity or chaos, but still being possible to explain or understand the underlying principles that govern the systems. A pragmatic view needs to be proposed: given the symmetries of the parts and the hierarchical laws of their interactions, it could be possible to extract the properties of the whole.

As an example, the multiscale properties of the tooth (Fig. 4.3) are analyzed with the bi-modal TR-NEWS imaging system [85]. The associated symmetrized nonlinear signal processing using TR invariance, reciprocity, and optimized excitations such as PI and ESAM [28], chirp-coded, or bi-solitonic signals has been included to standard NDT analysis [86]. Now, several experimental tests associated with advanced ultrasonic instrumentations are conducted in several laboratories around the world. Despite its original objective for industrial applications devoted to complex materials like composite [4], it opens also new perspectives in the context of medical imaging of complex media such as human tooth, bones, skin and, more ambitiously, the human brain evolution and its diseases.

Universality which consists in considering occurrences of structures or mechanism (protocol or politics in governing) in a great variety of systems, and diversity which consists in considering occurrences of structures or mechanism in many situations, commonly linked to a particular functionality, should be the keystone of the new development of a strategic governing of a complex and multiscale structure. This strategy will probably induce innovation in the design of self-adaptive structure and autonomous systems. The consideration of diverse hierarchical structures enables the development of functional sub-groups unique to a particular system despite the presence of few universal building groups (diversity/universality paradigm). The design of self-adaptive structure and autonomous systems of engineering is the natural consequence of such strategy: the design of self-replicating systems is based on the mother system, self-regenerating and self-healing enterprises and companies.

4.5   The Systemic Gold Proportion of a Complex System: Determinism Versus Stochasticity Evaluation

4.5.1   Multiscale Analysis and Stochastic Aspects in NDT 4.0

As shown above in the specific area of medical imaging, the increase of structure complexity is observed in all areas such as aeronautics or the nuclear power plant industry and its associated NDT and NIT. Furthermore, there is, in parallel, a need in the increase of accuracy of physical and technical parameters describing the complex structure under test, or under certification process. These two antinomical effects should be considered seriously with the same degree of interest, inducing the evidence of considering the problem with stochastic and multiscale analysis, and phenomenological approaches for simplicity of applicability. The systemic NDT optimal design considering these requirements should be generalized in the industry in the sense that new signal processing approaches are ready to be implemented in complex automated devices using suitable artificial intelligence. NDT engineering is going through a radical change of its intrinsic complexity, even though such a trend is far from being accepted by the community. Stochastics involving uncertainty in complex systems is a positive feature that should be considered as an openness of these systemic approaches, for modern industry 4.0 or NDT 4.0 [87].

The systemic NDT optimal design should be at the skeleton of NDT 4.0, the key component of the smart factory described in the emerging Industry 4.0 [88]. It consists in producing methods and systems capable of adaptation and robustness, no longer systems defined by stability and control. The robustness of modern NDT 4.0 methods should consider the ability of the method to maintain specific features when facing complex environment with several internal and external parameters, including human factor [89]. The systemic NDT 4.0 optimal design is also defined by the capability of developing the following attributes: resilience, adaptation, robustness, and scalability involving multiscality and multimodality. Regardless of how the concept of system is defined or specified by the different paradigms and approaches, the notion of complexity is based on the idea that information coming from such NDT complex systems could be accurately known (determinist) or could contain uncertainty (stochasticity). The degree of complexity could be estimated as the proportion of the stochastic part with respect to the deterministic one. This stochastic proportion should measure the multiscale property of any system, and constitutes the skeleton of an artificial intelligence associated with NDT 4.0.

This new information coming from the ratio between Stochasticity to determinism ratio (SDR) needs to be evaluated for all complex systems. The SNR for electronic measurements is also derived through SDR for generalized complex organizations. The first (linear) approach consisted in considering that the main deterministic part is the result of a linear superposition of other deterministic sub-parts, coming also from other sub-sub-parts; all this multiscale approach being limited by the bandwidth of the measurement system. This analysis induces finally a measurement of the information coming from the system which is generally affected by a stochastic information well-known as the noise, and being hopefully negligible with respect to the calibrated deterministic quantity. However, as explained in introduction, the stochastic part is naturally increased when the complexity of the system is increased too, and should be taken into account. The stochastic part induces consequently uncertainty. Nevertheless, uncertainty is a pregnant concept in the science of complexity and a promising future in the sense that uncertainty should be included in the study of complex systems for analyzing it with robustness.

In the future, NDT will not only be focused on "just testing," NDT 4.0 will be sorting, characterizing, monitoring, and checking as well, imitating the human brain that uses a multimodal and multiscale sensory system to extract information from noise and/or adapt its responses to the complex stochastic environment. As a result, NDT 4.0 has to develop cognitive and self-adapting sensor systems that will be able to decide for themselves what they measure when, where, and how, along with monitoring and characterization processes, etc. That is the standard we ought to be aiming for in NDT 4.0.

4.5.2 Dynamic of Nonlinear Systems: The Elementary Stochastic Cell of the Complexity

Nonlinear dynamics coming from nonlinear physics is nowadays introduced in many domains in science and engineering. During the last decades, many results have been found, and industrial applications concern most of modern systems and devices present in our society. Furthermore, the global way of thinking include now the nonlinear behavior as an evidence in terms of consequences. The example of the butterfly effect is now well understood and universally accepted as a nontrivial but possible event. In order to understand the concept of elementary cell of the complexity, the approach can be compared to kernel principal component analysis [90]. Each nonlinear system is defined in order to perform dimensionality reduction where linear methods were insufficient for identifying nonlinear signature in the data. The results of this linear dimensionality reduction are visualizations of the high dimensional data in a lower dimensional space (the elementary stochastic cell) that makes it possible to uncover patterns within the data using the nonlinear signatures of the resulting dynamic in the low dimensional space.

Multiplicity in initial states implies that a system can be very insensitive to changes in system input over a range of different hierarchical levels of the input, but once the level reaches a certain threshold the system is transformed into a completely different state with a modified hierarchy. The effect of nonlinearities is that a system future behavior might be difficult or impossible to predict, even in principle. The system future behavior is undetermined and analyses of such systems are therefore challenging. A concrete example of this complexity related to fluctuations is the observation of $1/f$ noise [25]. According to the measurements, some dynamical systems, many of them existing in nature, are organized into critical states where only a minor effect (in the probability sense) can cause disastrous consequences. The resultant consequences can be described as a $1/f$ power-law distribution, which is characteristic of the density spectrum. This power-law distribution is characterized by a heavy-tail, where the probability of extremely large effects cannot be neglected. According to this ubiquitous power law property, many systems are designed to be extremely robust against known perturbation, i.e., the ones that are known at the time of the design. The downside is that the system can become extremely fragile against perturbations it was not planned for to handle. The point to make here, however, is only that for many systems of interest in the present paper, the severity of the effects does not necessarily have to be linearly proportional to the magnitude of the causes.

4.5.3 Multiscality of Complex Systems: The Elementary Determinism Pattern of the Complexity

The structural complexity of these concepts is intrinsically associated with the increase of the complexity of experimental setup, or investigation protocol for social sciences, for example. The multiscale aspect of complex nonlinear systems is also present in synchronization processes. Some interesting regular behaviors were experimentally confirmed with the observation of invariants intrinsically identified using concepts coming from number theory [25]. This global property of synchronization is associated with a local one by the study of smaller elementary cells under simple nonlinear interaction.

Such analysis was proved with an experimental verification of the increase of fluctuations which associates $1/f$ noise to nonlinearity, autosimilarity, and regularity, represented here by the invariance property of the synchronization. In optical experiments, $1/f$ noise also appears in nonlinear multistable systems having many attractors with fractal properties. The $1/f$ noise resulted from interaction between attractors that are stable states of nonlinear dynamics. Synchronization zones are examples of stable states or invariant of nonlinear systems. Oscillators also exhibit multiscale characteristics (Fig. 4.7) of the synchronization involving arithmetical rules [91, 92]. For nonlinear measurements, the necessity to reach and acquire which was called in the past "hidden information in the noise" has conducted to think about calibrated advanced electronic systems as it was done for the ubiquitous $1/f$ noise previously described [25]. Generally, the experimental setup used for analyzing this complexity of $1/f$ noise (see Fig. 4.7) is designed using multiscale mixing electronic devices involving complex phase locked loops (PLL) cells. These mixing properties (Fig. 4.2) are the elementary cells of any measurements of a complex nonlinear systems [68]. This multiscale properties coming from the $1/f$ behavior has been also observed during the DNA denaturation process around the temperature transition [93]: the collective behavior of DNA bases being responsible of these fluctuation properties and DNA dissociation being at the

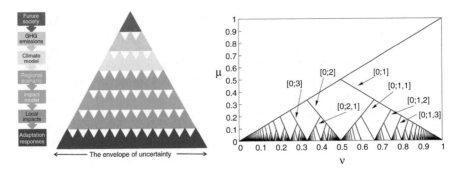

Fig. 4.7 A generic approach of complexity and vulnerability with their multiscale symmetries generated by arithmetical rules

origin of life evolution and the memory effect associated with genetic mutation. The link between memory effects and multiscale effects is again present in this case, as described in Part I.

Most of modeling of complex medium such as micro-inhomogeneized materials have shown some memory properties induced by the nonclassical nonlinearity [79] coming from structure degradation due to cracks [94], weak bonding, etc. In nonlinear NDT, the modeling of the hysteretic nonlinearity is based on the multiscale Preisach–Mayergoyz approach (PM space), as represented in Fig. 4.5. In this PM model, no analytical expression of the bulk modulus is given. It is calculated by summation of the strain contribution of a large number of elementary hysteretic elements (elementary cells). Each of these hysteretic element unit is described by two characteristic pore pressures corresponding to the transition between two states when the pore pressure is increased or decreased. The implementation of the PM space model is based on the multiscale approach. For each cell of the calculation grid (representing a mesoscopic level of the medium description), hysteretic units are considered with different values of the two characteristic stresses. This representation is commonly termed PM space and can be described mathematically by its density distribution using statistical decision tools [95]. In these examples, the multiscale properties should be associated with nonlinearity in order to take into account some memory properties in this complex system (Fig. 4.4). Again, two ingredients are mandatory: nonlinearity and multiscale properties.

4.6 Determinism Versus Stochasticity: A Pragmatic Approach for Engineering Organization's Strategies

Considering all the experiments and results described above, the consideration of nonlinear aspects in any system increases the complexity of any analysis. Furthermore, the wish to take into account the complexity of any system induces the need

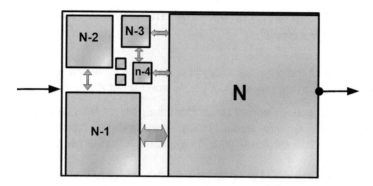

Fig. 4.8 The multiscale approach of a hierarchically connected system. If another cells ($n - 1$, $n - 2$, $n - 3$, etc.) are in multiscale interactions with the master cell N, then the iterative phenomenological scheme induces a gold proportion coming from gold number between deterministic information coming from the ($n, n - 1$) interaction ($\rho - 1 = 62\%$) and stochastic information coming from other (complex) interactions from other cells (38%). Any information coming as an output of system (represented by master cell N) contains a major deterministic part (62%) which is logically representative of the system and a minor stochastic part (38%) which comes from other "non-classical" paths

to use the tools developed and presented above. In this case, the knowledge of this complexity forces scientists, engineers, and social researchers to include a greater part of stochasticity as it was done previously: this stochasticity coming from the difficult knowledge to predict the deterministic part of complexity. Consequently, a phenomenological approach (like those used for PM space analysis described previously) could be extended to the evaluation of the multiscale property of a hierarchically connected system (Fig. 4.8) where input/output analysis is performed.

Let us try to describe and analyze the reasons which are at the origin that sometimes, a decision coming from a complex and structured organization is in contrary with the internal rules which are supposed to be followed. On the other side, it is frequently observed that the result of an action (the input) applied to a complex system induces a response (the output) which is on the contrary of any prediction made on this system. We are typically in the case where the severity of effects does not have a linear proportion with the magnitude of the causes. Surprisingly, we can find in our societies several examples of such nonsense: political elections, student's jury decisions, administrative board resolutions, champion's league, and any sports results. Even if these nonsense events constitute a small "stochastic" part with respect to the "deterministic" events, the complexity of the sub-elements that constitute the system needs to be taken into account in order to propose a phenomenological modeling of such "nonsense" responses in complex systems.

4.6.1 Excitation and Responses of Hierarchically Multiscale Complex Systems

The basic steps of the phenomenological construction described previously [10] is to assume the following properties of the system (Fig. 4.8):

1. The system contains a part of stochasticity, i.e., it is not fully deterministic.
2. The system, in its autonomous state, tries to find itself the best configuration with respect to the stochastic part: the autonomous state needs to be reached at a certain level.
3. The system is composed of several sub-systems which follow the same objective of robustness at different scales. The robustness is one of the multiscale properties that should be reached at all scales, but certainly with different properties.
4. At least one sub-system develops a nonlinear behavior. For example, threshold, saturation, and memory effects should be observed independently of the state of the system.
5. There is a hierarchical law between the sub-systems in the sense that actions performed to the whole system is not symmetrically distributed to the system, but hierarchically connected. On the contrary, the response given by the system comes hierarchically from sub-systems. The degree of hierarchy is increasing with respect to the complexity of the system, starting completely symmetrical for elementary cells.

Indeed, the multiscale approach of a hierarchically connected system can be useful for evaluating the deterministic/stochasticity ratio SDR defined previously. The approach of complexity could start assuming that there exists an elementary cell of complexity which contains the highest symmetry. If we assume that a system is simply characterized by the single subsystem I (Fig. 4.9a), one has the classical input/output relation with the system represented by a square. If we consider that the system contains two ($N = 2$) sub-systems (cells) with the same properties having symmetric rules, two squares with different size could represent the two-element subsystem (Fig. 4.9b). After considering that the length of the side $[a, b]$ represents the interaction between the two sub-systems X and Y, the optimal, stable, and robust configuration is obtained with the symmetric configuration. Consequently, the complexity could start from two elementary cells (X and Y) supposed to be identical and symmetrically connected. This hypothesis is of course well adapted for human interactions too, symbolized by letters representing chromosomes. Consequently, the optimal response coming from a system modeled with two sub-systems X and Y is those coming from 50% from X and 50% from Y. This initial state presents the advantage to be symmetric. This simple system with two degrees of freedom can be considered as the last case of a non-chaotic system inducing full deterministic and robust behavior.

The analysis of the first complex system is the 3-body system containing $N = 3$ hierarchically connected sub-systems. If we assume that the resulting system should

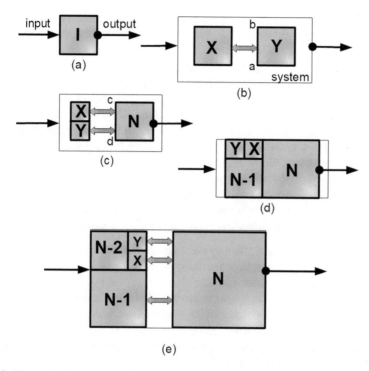

Fig. 4.9 The multiscale approach of a hierarchically connected system. The complexity modeling starts from two elementary cells (X and Y) supposed to be identical and symmetrically connected, and finishes with level N at the head of the system. (**a**) $N = 1$, (**b**) $N = 2$, (**c**) $N = 3$, (**d**) $N = 4$, (**e**) $N = 5$

contain the previous two-element case and preserve the hierarchical properties, the connection between the three sub-systems is given by Fig. 4.9c. If such system is excited by an external input, the response is given through the highest sub-system N, in the sense of the hierarchy supposed as an internal rule. It means that the response will be given by sub-system N taking into account (with a symmetrical ratio) both X and Y internal responses. The difference between the previous states is that the final decision will also contain its own response. This configuration maintains the symmetric interaction between cell N and lower cells X and Y with half of size $[c, d]$ for each of them, and preserves a kind of the symmetry between the three sub-systems N, X, and Y. In social organized systems with such configuration where N is the greatest element, any information coming from N (the manager) will come from the same symmetric interactions between sub-cells X and Y. Note that the length of side $[c, d]$ is doubled taking into account that the greatest element N should contain itself additional connections represented by a greater square size. This greater size could represent a kind of superiority of element N with respect to elements X and Y, giving the starting point of the hierarchical property. This metric relation associated with a distance (or a norm in the mathematical sense) has several equivalent for pragmatic situations: age, experience in social sciences, knowledge, etc.

For $N = 4$, $N = 5$, etc., the same iterative process could be proposed (as shown in Fig. 4.9d and e). In the four-element hierarchically connected case, cell N is supposed to provide the main "representability" of the whole system. Any information coming from N comes hierarchically from interaction with cell $N - 1$ (50%), and interactions with cell X or Y, in this 2D representation case. This iterative process can be extended as the complexity increases and induces interesting conclusions. Some basic and phenomenological pragmatic rules can be extracted from this systemic analysis of any complex system structured hierarchically (Fig. 4.8) leading to the golden number $\rho = 1.618$. If a hierarchical property is supposed in the complex system, some interesting behaviors could be extracted following this multiscale approach of this engineering organization's strategy:

1. $\rho - 1 = 62\%$ of decision taken by sub-system N is coming from the sub-system $N - 1$ (in the optimal multiscale limit). This is a guarantee of the confidence between N and $N - 1$ interactions.
2. The simplest case of a complex system contains two elementary cells (X and Y in Fig. 4.9) with a hierarchical cell ($N - 4$ in Fig. 4.8) is logically characterized by the fact that half of decision comes from cell Y and half comes from cell X. This proportion is schematically measured by the common side (or interaction) of the square associated with the cell.
3. A non-negligible part of decision taken by sub-system N could be given by interaction between sub-system N and sub-system $N - 4$ for example (Fig. 4.4), which could have a small interaction (minority). This induces a potentiality that minority could take part of strategic decisions at a high level of organized structures, shunting the "logical" hierarchy between N and $N - 1$. This consequence is interesting in the sense that it allows the possibility of a dynamical property between cells N_i. If cell $N - 4$ is at the origin of a positive decision (not validated by cell $N - 3$), cell $N - 2$ can decide to permute cell positions $N - 4$ and $N - 3$. This can be seen as a kind of recognition leading to an increase in the hierarchical level, or a dynamical hierarchy.
4. Any elementary cell in increasing evolution should satisfy the proportion of 62% of the internal rules governing the sub-system where it is included. The 38% complement is the result (or the objective!) of an opposite rule being against the internal rules of the sub-system.
5. Proportions of 100% and 0% constitute some extreme conditions leading to, respectively, totally deterministic behavior and totally stochastic behaviors of the whole system.
6. The multiscale properties of the whole system is a fundamental key of this proportion. Multiscality should be present in the excitation, the modeling, and the analysis of the system. Without any limitation in the scale of the analysis, the proportion between N and $N - 1$ interaction and between N and $\sum(N - i)$ should reach the golden number ρ as shown by the classical Fig. 4.8.

7. Multiscality of a complex system containing multiscale sub-systems where the previous rules are applied is a guarantee of the optimal robustness of the whole system itself. Any breaking of this multiscale symmetry will lead to global instability.

This approach could be pragmatically applied to engineering organization's strategies for innovation and social monitoring:

1. $\rho - 1 = 62\%$ of decided actions should follow the global strategy of any organization (the majority)
2. $2 - \rho = 38\%$ of decided actions should be against the global strategy of any organization (the minority)

Consequently, decided actions that follow strategic decisions (or decisions approved by the majority) is a guarantee of the stability (or the legitimacy) of any organization or social community. Decided actions that are against strategic decisions is a guarantee of the innovative potentiality and creativity of any organization. These actions against the "evidence" is the basis for any "disruption" or breakthrough observed in our societies. Any 0% in one of these parts will conduct to extreme behavior known as totalitarianism and anarchy in political organizations.

Finally, since the basic output (decision) of a democratic system is presently conditioned by the majority, this hierarchically multiscale approach could improve any democratic system by considering some decisions coming from minority (Fig. 4.10), which are usually responsible of large actions in order to force the "democratic" system to include their point of view [10, 97]. The multiscale pragmatic probability law could be also used in order to take into account the protest vote (or the "blank vote"). For example, a single candidate obtaining 51.01% of positive votes and 49.99% of "blank vote" would only have $\rho - 1 = 62\%$ of probability of success (Fig. 4.10). Conducted at all scales of any hierarchical structured organization, this could be a proposal to avoid the well-known concept of the tyranny of the majority, reported few centuries ago by Alexis de Toqueville [98].

Fig. 4.10 The multiscale pragmatic probability law. Any democratic decision taken with a proportion of 50.01% should be applied only with a probability of $\rho - 1 = 62\%$. If this rule is applied at all scales of subsystems, the multiscale pragmatic probability law is given by the Devil's staircase also observed in complex systems [92, 96]

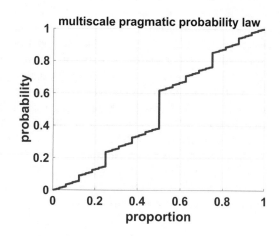

4.6.2 Mutability, Tunability, Optimality, Evolvability Coming from the Nano-scale Science

Multiscale analysis is consequently necessary to our understanding of how complex engineering and social systems rules contribute to a better implementation of innovation of strategic governing and entrepreneurship. Hierarchical structure should be identified with the different levels of connections. Such complex and also nonlinear behavior like mutability, tunability, optimality, evolvability, universality, and diversity is coming from the nanoscale science. It should be applied practically to define the new engineering entrepreneur with the objective to propose new strategic innovations and optimized entrepreneurship for the engineering of complex systems and complex structures, human and social organizations, clearly compatible with the fourth industrial revolution Industry 4.0 [87].

For example, two opposite aspects like mutability (the capacity to change functional properties of a system based on external excitations) and tunability (the capacity to change the system reversibly during use) should be taken into account at all scales of the system [99]. Both should be clearly considered for strategic governing. The adaptation to reach a desired characteristic while respecting a set of restrictions in the excitation (optimality) should also be included and put in balance hierarchically with the evolvability, which is the ability to acquire new functions or features in response to changed excitations.

4.6.3 Vulnerability of Stochastic Data and Cascade of Uncertainty

The present part concerns methods and knowledge that are useful when analyzing the risks and vulnerabilities of complex systems in a societal emergency management context. Operational definitions of vulnerability and emergency response capabilities are suggested and two methods for analyzing the vulnerability of critical infrastructure networks, based on the suggested definition, are presented. An empirical study of people values and preferences regarding different attributes of potential disaster scenarios is also presented, since knowledge about values also is crucial for adequate risk and emergency management.

Most of studies conducted are done with judgments under certainty. However, when the cell N is making decisions regarding future possible risk scenarios, one cannot know for sure which scenario will occur. Therefore, it should be interesting to study how the other values (coming from cells $N - i$) and preferences for the attributes would change if the trade-offs are framed in terms of judgments under uncertainty. This is done for example in the context of climatology. Explicitly modeling the SDR in making decisions enables adequate recommendations to be drawn regarding the effect of these uncertainties on the political decision made.

4.6.4 Multilevel Approach in Social and Epidemiological Sciences

Because many technological systems, political decisions today are complex systems, understanding their susceptibility to collective failure is a critical problem. Understanding vulnerability in complex systems requires an approach that characterizes the coupled interactions at multiple scales of cascading failures [100]. It is important to develop a multiscale processing pipeline that can characterize the vulnerability of complex systems in order to improve the evidence of its presence in any systems. The vulnerability of any information could help anyone to accept errors, suggest forgiveness, and develop pragmatism. Spatiotemporal analysis and dimensional reduction techniques are the keystones in order to find symmetries of the system leading to collective behavior and extreme events.

This multilevel signal processing is designed to explore and analyze data that come from populations which have a complex structure. In any complex structure we can identify atomic units, like HEU defined in the PM-space approach (see Fig. 4.5). These are the units at the lowest level of the system (Fig. 4.11). The response of each y variable is measured on the atomic units. Often, but not always, these atomic units are individuals. Individuals are then grouped into higher level units, for example, neighborhoods. By convention, we say that individuals are at level 1, households are at the level 2, neighborhoods are at level 3 in our structure, etc. A level (e.g., pupils, schools, households, areas) is made up of a number of individuals units (e.g., particular pupils, schools, etc.). The term level can be used somewhat interchangeably. Nevertheless, the term level implies a nested hierarchical relationship of units (in which lower units nest in one, and one only, higher level unit), whereas classification does not. Since the global hierarchy is

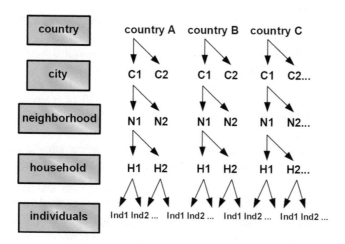

Fig. 4.11 Hierarchical structure of social data into levels that are connected nonlinearly with local rules

defined, local rules between levels should be defined with a systemic and pragmatic way. It means that some basic elementary components such as threshold, limits, and proportionality should be preferred for local interactions, including of course nonlinear responses.

4.7 Conclusion

As a pragmatic conclusion, if we consider any complex system as a hierarchically structured system with multiscale three- or four-parameters simple nonlinear local rules, the presented tools of big data processing can be used for defining the dynamic of the complexity. This paper provides important insight into some of the issues related to the methods and materials for monitoring risks, vulnerabilities, and integrity of complex systems; investigated in NDT and biomedical imaging areas. In order to generalize these methods in all domains of sciences, more research is needed to address this difficult and comprehensive task. A real interaction should be established between scientists from engineering sciences that are applying nonlinear tools and results, and researchers from social sciences that are measuring effects in our complex society. In modern NDT or NDT 4.0, such approaches are now stated including human factors [89].

Concerning our research in this domain, generalized multiscale nonlinear innovations have been presented, including advanced ground truth multimodal applications:

- New perspectives for multimodal medical imaging and NDT in complex samples.
- TR-NEWS is presented as a progressive method, which uses energy focusing during the time-reversal process, and the influence of nonlinearity on the TR system response.
- Connection of TR techniques represents powerful technique for extraction of nonlinear signatures in a medium using symmetry properties.
- Improvement of TR-NEWS sensitivity with optimized excitations is validated in order to activate nonlinear signature and memory properties.
- More generally, nonlinear time reversal based methods can have new applications in many areas of research: non-Invasive testing (NIT) of biomaterials (bone, skin, tooth, human brain, etc.), cryptography, non-destructive testing (NDT) of complex media, etc.
- Symmetry analysis including theoretical and mathematical tools like Lie groups is necessary for understanding the structural skeleton of the complexity.

A generic process for studying aging of complex systems is also described with the objective to be applicable practically in the domain of modern multimodal medical imaging and in the industrial domain of non-destructive testing. Both of these areas where academic sciences such as modern mathematics, nonlinear physics, modern technology, and stochastic aspects of human sciences should be introduced in order to evaluate disruptive potential applications of big data sciences and artificial intelligence. Starting from the deterministic modeling of material sciences, the linear ultrasonic characterization of materials and biological medium is improved in order to take into account the evidence of inhomogeneity, nonstationarity, and nonlinearity of any complex medium under analysis. From the technical point of view, such an evidence induced a complete optimization of the instrumentation, including modern numerical devices, high speed microprocessors, multimodal and multiphysics sensors and actuators associated with an accurate metrology and noise analysis, and a huge capacity for the big data storage and digital processing processors. From the theoretical point of view, the huge area of linear physics associated with stationarity and deterministic systems is completed with the increasing area of nonlinear and stochastics new tools. The results are in the extraction of a new family of multiscale data, considered as undesired noise 20 years ago, but now enhanced as potential big data inputs, having the intrinsic signature of new statistical properties like memory, nonstationarity, ergodic, chaotic signatures. These new data are considered as strong candidates for accurate evaluation of aging processes, including the conservation and rehabilitation of pieces of the historical and cultural heritage like stones, paintings, jewelry, or any objects fabricated by our ancient civilization.

This review paper has the objective to convince researchers, engineers, professionals, physicians, and end-users to follow the latest developments experienced by different methods in ultrasonic experimentations, including the use of modern mathematical models. The objective is to extend the concept of "optimization of excitations" by recent researches with applications in several domains (electrical engineering, mechanical engineering, computer engineering, industrial engineering, telecommunications engineering, finance, social sciences, etc.) by the various multiscale and multimodal imaging and signal processing tools, including of course some perspectives with the use of artificial intelligence, the merged disciplines of science, technology, engineering, and mathematics.

Acknowledgements This work has been partially supported by the Région Centre-Val de Loire (France) under the PLET project 2013-00083147. I would like to thank Prof. Kasso A. Okoudjou and Prof. Sadataka Furui for valuable discussions related to mathematical aspects of TR-NEWS concepts. Prof. Mostafa Maslouhi is also indebted for his invitation to present these ideas during the TEM2018 conference held in Kenitra, Morocco, March 2018.

References

1. B.E. Anderson, M.C. Remillieux, P.Y. Le Bas, T. Ulrich, in *Nonlinear Ultrasonic and Vibro-Acoustical Techniques for Nondestructive Evaluation* (Springer, 2019), pp. 547–581
2. G. Busse, D. Van Hemelrijck, I. Solodov, A. Anastasopoulos, *Emerging Technologies in NDT*. Balkema - proceedings and monographs in engineering, water and earth sciences (Taylor & Francis, 2008). https://books.google.fr/books?id=M0KEYUaJuDMC
3. D. Burgos, L. Mujica, J. Rodellar, *Emerging Design Solutions in Structural Health Monitoring Systems*. Advances in Civil and Industrial Engineering Series (IGI Global, 2015)
4. P. Duchene, S. Chaki, A. Ayadi, P. Krawczak, Journal of Materials Science pp. 1–24 (2018)

5. S. Dos Santos, A. Masood, S. Furui, G. Nardoni, in *2018 16th Biennial Baltic Electronics Conference (BEC)* (2018), pp. 1–4. https://doi.org/10.1109/BEC.2018.8600977
6. J. Moysan, G. Corneloup, T. Sollier, NDT & E International **32**(2), 79 (1999)
7. M. Sezgin, B. Sankur, in *Image Processing, 2001. Proceedings. 2001 International Conference on*, vol. 3 (IEEE, 2001), vol. 3, pp. 764–767
8. D. Mery, V. Riffo, U. Zscherpel, G. Mondragón, I. Lillo, I. Zuccar, H. Lobel, M. Carrasco, Journal of Nondestructive Evaluation **34**(4), 42 (2015)
9. Z. Farova, V. Kus, S. Dos Santos, in *Proc of the Forum Acusticum* (Aalborg, 2011), pp. 991–996
10. S. Dos Santos, in *Proceedings of the SPMS Stochastic and Physical Monitoring Systems International Conference* (2012), ISBN 978-80-01-05130-6, pp. 19–40
11. C. Chen, *Signal and Image Processing for Remote Sensing, Second Edition*. Electrical engineering/remote sensing (Taylor & Francis, 2012). https://books.google.fr/books?id=QQDHl3L867QC
12. S. Dos Santos, S. Vejvodova, Z. Prevorovsky, Proceedings of the Estonian Academy of Sciences **59**, 301 (2010)
13. S. Dos Santos, N. Poirot, M. Brochard, in *2016 IEEE International Ultrasonics Symposium (IUS)* (2016), pp. 1–4. https://doi.org/10.1109/ULTSYM.2016.7728856
14. G. Filippidis, M. Mari, L. Kelegkouri, A. Philippidis, A. Selimis, K. Melessanaki, M. Sygletou, C. Fotakis, Microscopy and Microanalysis **21**(2), 510–517 (2015). https://doi.org/10.1017/S1431927614013580
15. S. Dos Santos, M. Lints, N. Poirot, A. Salupere, The Journal of the Acoustical Society of America **138**(3), 1796 (2015)
16. D. Broda, W. Staszewski, A. Martowicz, T. Uhl, V. Silberschmidt, Journal of Sound and Vibration **333**(4), 1097 (2014). https://doi.org/10.1016/j.jsv.2013.09.033. http://www.sciencedirect.com/science/article/pii/S0022460X13007876
17. G.P.M. Fierro, M. Meo, Ultrasonics **93**, 43 (2019)
18. K. Van Den Abeele, A. Sutin, J. Carmeliet, P.A. Johnson, NDT E International **34**, 239 (2001)
19. O. Bou Matar, S. Dos Santos, M. Vila, F. Vander Meulen, in *Ultrasonics Symposium, 2002. Proceedings. 2002 IEEE*, vol. 1 (IEEE, 2002), vol. 1, pp. 881–884
20. M. Vila, F. Vander Meulen, S. Dos Santos, L. Haumesser, O. Bou Matar, Ultrasonics **42**, 1061 (2004)
21. N. Li, J. Sun, J. Jiao, B. Wu, C. He, Ndt & E International **79**, 63 (2016)
22. F. Ciampa, S.G. Pickering, G. Scarselli, M. Meo, Structural Control and Health Monitoring **24**(5), e1911 (2017)
23. R.A. Guyer, P.A. Johnson, Physics Today **52**, 30 (1999)
24. R.A. Guyer, K.R. McCall, G.N. Boitnott, Phys. Rev. Lett. **74**, 3491 (1994)
25. M. Planat, S. Dos Santos, N. Ratier, J. Cresson, S. Perrine, in *Proc. of the VII Van der Ziel Symposium on Quantum 1/f Noise and other Low Frequency Fluctuations in Electronic Devices*, ed. by P.H. Handel, A.L. Chung (1999), p. 177
26. V. Sanchez-Morcillo, N. Jimenez, J. Chaline, A. Bouakaz, S. Dos Santos, in *Localized Excitations in Nonlinear Complex Systems, Nonlinear Systems and Complexity*, vol. 7, ed. by R. Carretero-Gonzalez, J. Cuevas-Maraver, D. Frantzeskakis, N. Karachalios, P. Kevrekidis, F. Palmero-Acebedo (Springer International Publishing, 2014), pp. 251–262
27. S. Dos Santos, N. Jimenez, V. Sanchez-Morcillo, in *2016 IEEE International Ultrasonics Symposium (IUS)* (2016), pp. 1–4. https://doi.org/10.1109/ULTSYM.2016.7728853
28. S. Dos Santos, C. Plag, Int. Journal of Non-Linear Mechanics **43**, 164 (2008)
29. B.J. Cantwell, *Introduction to Symmetry Analysis* (Cambridge University Press, 2002)
30. S. Dos Santos, B. Choi, A. Sutin, A. Sarvazyan, in *Proc. of the 8ème Congrès Francais d'Acoustique* (Tours, 2006), pp. 359–362
31. J. Chaline, N. Jimenez, A. Mehrem, A. Bouakaz, S. Dos Santos, V.S. Morcillo, The Journal of the Acoustical Society of America **138**(6), 3600 (2015). https://doi.org/10.1121/1.4936949. http://scitation.aip.org/content/asa/journal/jasa/138/6/10.1121/1.4936949
32. M. Frazier, B. Taddese, T. Antonsen, S.M. Anlage, Phys. Rev. Letters **110**, 063902 (2013)

33. J. Marotti, S. Heger, J. Tinschert, P. Tortamano, F. Chuembou, K. Radermacher, S. Wolfart, Oral and Maxillofacial Radiology **115**, 819 (2013)
34. M. Miniaci, A. Gliozzi, B. Morvan, A. Krushynska, F. Bosia, M. Scalerandi, N. Pugno, Physical review letters **118**(21), 214301 (2017)
35. T. Ulrich, A. Sutin, R. Guyer, P. Johnson, International Journal of Non-Linear Mechanics **43**(3), 209 (2008)
36. G. Zumpano, M. Meo, Int. J. Solids Struct. **44**, 3666 84 (2007)
37. A. Sutin, B. Libbey, L. Fillinger, L. Sarvazyan, J. Acoust. Soc. Am. **125**, 1906 (2009)
38. L.A. Ostrovsky, O.V. Rudenko, in *Proc. of the 18th International Symposium on Nonlinear Acoustics ISNA* (Stockholm, 2008), pp. 9–16
39. M. Lints, A. Salupere, S. Dos Santos, Proceedings of the Estonian Academy of Sciences **64**, 297 (2015). (IF2015: 0.794)
40. O. Yesiltas, Physica Scripta **80**(5), 055003 (2009)
41. P. Trampus, V. Krstelj, in *Proc of the ECNDT 2018* (2018)
42. L. Wenhui, J. Cheng, S. Dos Santos, Y. Chen, International Journal of Computational and Engineering **2**, 235 (2017)
43. J. Wu, Applied Mathematics and Computation **214**(1), 31 (2009). https://doi.org/10.1016/j.amc.2009.03.066. http://www.sciencedirect.com/science/article/pii/S0096300309002756
44. Y. Chen, Y. Wu, Y. Cui, Z. Wang, D. Jin, Journal of Computational Science **1**(3), 146 (2010). https://doi.org/10.1016/j.jocs.2010.07.001. http://www.sciencedirect.com/science/article/pii/S1877750310000426
45. D. Remache, M. Caliez, M. Gratton, S. Dos Santos, Journal of the mechanical behavior of biomedical materials **77**, 242 (2018). https://doi.org/10.1016/j.jmbbm.2017.09.009. http://www.sciencedirect.com/science/article/pii/S1751616117303922
46. F. Mainardi, *Fractional calculus and waves in linear viscoelasticity: an introduction to mathematical models* (World Scientific, 2010)
47. D. Craiem, F.J. Rojo, J.M. Atienza, R.L. Armentano, G.V. Guinea, Physics in Medicine & Biology **53**(17), 4543 (2008)
48. I.D. Mayergoyz, J. Appl. Phys. **57**, 3803 (1985)
49. S. Dos Santos, C. Kosena, V. Kus, D. Remache, J. Pittet, M. Gratton, M. Caliez, in *Proceedings of the 23nd International Congress on Sound and Vibration, Major challenges in Acoustics, Noise and Vibration Research.* (2016)
50. L. Chua, IEEE Transactions on circuit theory **18**(5), 507 (1971)
51. G. Indiveri, S.C. Liu, Proceedings of the IEEE **103**(8), 1379 (2015)
52. J.J. Yang, D.B. Strukov, D.R. Stewart, Nature nanotechnology **8**(1), 13 (2013)
53. A. Ascoli, R. Tetzlaff, F. Corinto, M. Mirchev, M. Gilli, in *2013 14th Latin American Test Workshop-LATW* (IEEE, 2013), pp. 1–6
54. F.Z. Wang, N. Helian, S. Wu, M.G. Lim, Y. Guo, M.A. Parker, IEEE Electron Device Letters **31**(7), 755 (2010)
55. G. Johnsen, C. Lütken, Ø. Martinsen, S. Grimnes, Physical Review E **83**(3), 031916 (2011)
56. S. Dos Santos, S. Furui, in *2016 IEEE International Ultrasonics Symposium (IUS)* (2016), pp. 1–4. https://doi.org/10.1109/ULTSYM.2016.7728885
57. S. Dos Santos, A. Masood, M. Lints, A. Salupere, C. Kozena, V. Kus, J.C. Pittet, M. Caliez, M. Gratton, in *25th International Congress on Sound and Vibration (ICSV25)* (2018)
58. H. Ghorbel-Feki, A. Masood, M. Caliez, M. Gratton, J.C. Pittet, M. Lints, S. Dos Santos, Comptes Rendus Mécanique **347**(3), 218 (2019)
59. C. Kozena, V. Kus, S. Dos Santos, in *2016 15th Biennial Baltic Electronics Conference (BEC)* (2016), pp. 179–182. https://doi.org/10.1109/BEC.2016.7743758
60. S. Kim, C. Du, P. Sheridan, W. Ma, S. Choi, W.D. Lu, Nano letters **15**(3), 2203 (2015)
61. M.A. Zidan, J.P. Strachan, W.D. Lu, Nature Electronics **1**(1), 22 (2018)
62. J. Papouskova, S. Dos Santos, V. Kus, in *in Proc. of the BEC international Baltic IEEE conference* (2012), IEEE Catalog Number CFP12BEC-PRT, ISSN 1736-3705, pp. 323–326
63. S. Dos Santos, Z. Prevorovsky, Ultrasonics **51**(6), 667 (2011)
64. G. Croain, R. Morel, Z. Prevorovský, S. Dos Santos, in SPMS 2013 proceedings (2013)

65. D. Chimura, R. Toh, S. Motooka, IEEE transactions on ultrasonics, ferroelectrics, and frequency control **64**(12), 1874 (2017)
66. S. Dos Santos, in *Proceedings of the Joint Congress CFA/DAGA'04* (Strasbourg, 2004), pp. 549–550
67. L. Schmerr, *Fundamentals of Ultrasonic Nondestructive Evaluation: A Modeling Approach*. Springer Series in Measurement Science and Technology (Springer International Publishing, 2016). https://books.google.fr/books?id=aNAYDAAAQBAJ
68. S. Dos Santos. The nonlinear mixing of waves: the up-and-coming method for transmission, evaluation and metrology. Academia NDT International invited Lecture (2011). http://www.academia-ndt.org/admin/Downloads/Topo_Academia-Munich2016-V2.pdf
69. C. Heaton, B.E. Anderson, S.M. Young, The Journal of the Acoustical Society of America **141**(2), 1084 (2017)
70. I. Petromichelakis, C. Tsogka, C. Panagiotopoulos, Wave Motion **79**, 23 (2018). https://doi.org/10.1016/j.wavemoti.2018.02.007. http://www.sciencedirect.com/science/article/pii/S0165212518300623
71. M. Lints, S. Dos Santos, A. Salupere, Wave Motion **71**, 101 (2017)
72. P. Hydon, *Symmetry methods for Differential Equations: A beginner's Guide* (Cambridge University Press, 2000)
73. B.J. Cantwell, *Introduction to Symmetry Analysis* (Cambridge University Press (612 pages), 2002)
74. A.D. Polyanin, V.F. Zaitsev, *Handbook of nonlinear partial differential equations* (Chapman and Hall/CRC, 2016)
75. J.G. Belinfante, B. Kolman, *A Survey of Lie Groups and Lie Algebra with Applications and Computational Methods*, vol. 2 (SIAM, 1989)
76. S. Dos Santos, J. Chaline, (American Institute of Physics Inc., 2015), vol. 1685. https://doi.org/10.1063/1.4934412
77. M. Moran, R. Gaggioli, SIAM J. Appl. Math. **16**, 202 (1968)
78. Y.B. Gaididei, J.F. Archilla, V.J. Sánchez-Morcillo, C. Gorria, Physical Review E **93**(6), 062227 (2016)
79. T. Goursolle, S. Callé, S. Dos Santos, O. Bou Matar, J. Acoust. Soc. Am. **122**(6), 3220 (2007)
80. S. Vejvodova, Z. Prevorovsky, S. Dos Santos, in *Proceedings of Meetings on Acoustics XIII-ICNEM*, vol. 3 (ASA, 2008), vol. 3, p. 045003
81. P. Blanloeuil, L.F. Rose, M. Veidt, C.H. Wang, Journal of Sound and Vibration **417**, 413 (2018)
82. F. Wang, G. Song, Mechanical Systems and Signal Processing **130**, 349 (2019)
83. V. Sanchez-Morcillo, N. Jiménez, S. Dos Santos, J. Chaline, A. Bouakaz, N. Gonzalez, Modelling in science education and learning **6**(3), 75 (2013)
84. D. Levi, M.A. Rodriguez, J. Phys. A: Math. Gen. **37**, 1711 (2004)
85. S. Dos Santos, M. Domenjoud, Z. Prevorovsky, Physics Procedia **3**(1), 913 (2010)
86. M. Lints, A. Salupere, S. Dos Santos, Acta Acustica united with Acustica **103**(6), 978 (2017)
87. S. Dos Santos, Z. Prevorovsky, C. Mattei, V. Vengrinovich, G. Nardoni, in Proc. of the ECNDT 2018 conference (2018). https://www.ndt.net/article/ecndt2018/papers/ecndt-0623-2018.pdf
88. S. Wang, J. Wan, D. Zhang, D. Li, C. Zhang, Computer Networks **101**, 158 (2016)
89. M. Bertovic, M. Gaal, C. Müller, B. Fahlbruch, Insight-Non-Destructive Testing and Condition Monitoring **53**(12), 673 (2011)
90. B. Schölkopf, A. Smola, K.R. Müller, in *Artificial Neural Networks — ICANN'97*, ed. by W. Gerstner, A. Germond, M. Hasler, J.D. Nicoud (Springer Berlin Heidelberg, Berlin, Heidelberg, 1997), pp. 583–588
91. S. Dos Santos, S. Dos Santos, in *Proceedings of the SPMS Stochastic and Physical Monitoring Systems International Conference* (2013), ISBN 978-80-01-05383-6, pp. 19–36
92. S. Furui, T. Takano, International Journal of Bifurcation and Chaos **25**(11), 1550145 (2015)
93. K.S. Nagapriya, A.K. Raychaudhuri, Phys. Rev. Lett. **96**, 038102 (2006)

94. S. Delrue, V. Aleshin, K. Truyaert, O.B. Matar, K. Van Den Abeele, Ultrasonics **82**, 19 (2018)
95. V. Kus, J. Papouskova, S. Dos Santos, in *Proc. of the 6th NDT in Progress 2011 International Workshop of NDT Experts* (2011), ISBN 978-80-214-4339, pp. 157–165
96. M. Planat, F. Lardet-Vieudrin, G. Martin, S. Dos Santos, G. Marianneau, Journal of applied physics **80**(4), 2509 (1996)
97. C. Wernz, A. Deshmukh, European Journal of Operational Research **202**(3), 828 (2010). https://doi.org/10.1016/j.ejor.2009.06.022. http://www.sciencedirect.com/science/article/pii/S0377221709004913
98. A. de Tocqueville, *De la démocratie en Amérique*. No. vol. 1 in De la démocratie en Amérique (M. Lévy frères, 1868). https://books.google.fr/books?id=ShxSvVvbnaQC
99. T.P.J. Knowles, M.J. Buehler, Nature Nanotechnology **6**, 469 (2011)
100. V. Misra, D. Harmon, Y. Bar-Yam, in *SSS 2010, LNCS*, ed. by S.D. et al. (Springer-Verlag Berlin Heidelberg, 2010)

Chapter 5
A Geometrically Constrained Manifold Embedding for an Extrinsic Gaussian Process

Thomas Deregnaucourt, Chafik Samir, Abdelmoujib Elkhoumri, Jalal Laassiri, and Youssef Fakhri

Abstract We introduce a new framework of local and adaptive manifold embedding for Gaussian regression. The proposed method, which can be generalized on any bounded domain in $\mathbb{R}^n$, is used to construct a smooth vector field from line integral on curves. We prove that optimizing the local shapes from data set leads to a good representation of the generator of a continuous Markov process, which converges in the limit of large data. We explicitly show that the properties of the operator with respect to a geometry are influenced by the constraints and the properties of the covariance function. In this way, we make use of Markov fields to solve a registration problem and place them in a geometric framework. Finally, this locally adaptive embedding can be used with the help of the linear operator to construct conformal mappings or even global diffeomorphisms.

Keywords Vector field · Gaussian process · Random field · Covariance operator

5.1 Introduction

Statistical analysis and modeling of shapes of objects take their origin in works established by Kendall in 1984 [1] where the shape of an object in Euclidean space is defined as all the geometrical information that remains when location, scale, and rotational effects are filtered out from an object. Many different methods for fitting geodesics in Kendall shape space have been proposed. While Kendall's definition of shape space took major strides in shape analysis, it admits some limitations due

T. Deregnaucourt · C. Samir (✉)
CNRS-UCA UMR 5961, Clermont-Ferrand, France
e-mail: Thomas.Deregnaucourtr@uca.fr; chafik.samir@uca.fr

A. Elkhoumri · J. Laassiri · Y. Fakhri
Ibn Tofail University, Kenitra, Morocco
e-mail: laassiri@uit.ac.ma; fakhri@uit.ac.ma

© Springer Nature Switzerland AG 2020
S. Dos Santos et al. (eds.), *Recent Advances in Mathematics and Technology*,
Applied and Numerical Harmonic Analysis,
https://doi.org/10.1007/978-3-030-35202-8_5

to the use of landmarks to define shape space. Therefore, much work has been done in order to find a convenient representation of shapes that enables simple physical interpretations of deformations of shapes and efficient method for fitting curves and geodesics. Klassen et al. [2–5] propose a new geometric representation of curves based on computational differential geometry. First, they were restricted to arc-length parametrization of curves [2]. Therefore, shapes are represented as elements of infinite-dimensional spaces. Then, they propose a square-root velocity (SRV) representation for analyzing shapes of curves in Euclidean spaces under an elastic metric and compute geodesics between closed curves using path-straightening approach. Other authors have also presented other variational techniques for finding diffeomorphic deformation vector fields between curves [6, 7, 13].

The recent advances in imaging have led to an increased need for image registration methods which are used in a large number of applications including medical imaging, computer vision, graphics, etc. The image registration problem consists of mapping a target image to a reference image under certain constraints. The estimated deformation can be based on intensity (gray-scale level correspondences), geometry (features or landmarks), or both. The registration problem can be rephrased as a variational or statistical problem where several different models are available to predict the deformation vector field on the whole image domain. Thus, one has to build an efficient model that best matches the given landmarks (points, curves, surfaces, etc.) accurately, e.g., been smooth enough on the rest of the domain [14].

Generally, one looks for the deformation field that best maps one image (or a part of it) onto another one. This is a classical variational (ill-posed) problem, which is usually solved by adding a regularization term [11]. Thin plate spline image registration [15] is the standard method for matching points under the assumption that the point-wise and full deformations are small. For example, in [12], the deformation field was driven by a minimization flow toward a harmonic map corresponding to the solution of a coupling of data and regularization terms. For large deformations, a diffeomorphic matching approach was developed by Grenander et al. [13]; it was followed by other deformable template-based approaches [8, 9].

Landmark-based image registration is based on finite sets of landmarks, usually not uniformly distributed, where each landmark in the source image has to be mapped onto the corresponding landmark in the target image. The landmark-based registration problem can be formulated in the context of multivariate random fields, and solved by different numerical methods [10]. Despite its popularity, two of its main issues are sensitivity to landmark locations and dependency on point correspondences. Recently, several methods have been proposed to deal with these disadvantages. These statistical techniques, giving rise to compactly supported or local mappings, handle locally deformed images. Moreover, they are generally stable and the computational effort to determine transformations is low. Thus, they can deal with a large number of landmarks.

For large deformations, a diffeomorphic matching approach was developed by Grenander et al. [13]; it was followed by other deformable template-based approaches such as the large deformation diffeomorphic metric mapping (LDDMM) method to solve for large deformations when the landmark correspondences are known [9]. Other approaches can be found, where the deformation fields are defined on smooth regions with distinct "anatomical" landmark points. In this respect, geometric regularization and stochastic formulations offer some nice advantages. Specifically, deformable image registration can be formulated as a stochastic optimization problem, in which the likelihood term is coupled with regularization prior to ensure smooth solutions. In this work, geometric landmarks are given as curves without any correspondences between points (correspondences are only given at the curve level). In this setting, previous methods are not directly applicable, and one has to develop a unified framework that can efficiently compute point correspondences across curves and the full deformation fields jointly. This motivates us to focus on curve-based diffeomorphic image registration, and to develop a novel probabilistic model that encodes shape variability of the landmark curves [17].

This paper proposes a new framework to solve point correspondences across curves using shape analysis and the full deformation fields jointly using a Gaussian process. Gaussian processes (GP) or Markov random fields are a state-of-the-art probabilistic non-parametric regression method. In order to capture a smooth deformation vector field between two observed set of curves, build a probabilistic model, and perform optimal predictions for non-observed data on the image domain Ω, we aim to study a GP as a distribution over the geodesic deformation field U between landmark curves. In fact, $U \sim GP(\mu, C)$ on Ω and is fully defined by a mean function μ (in our case $\mu = 0$) and a covariance function C. To reach such goal we will present properties of the Gaussian process. In this work, we will first show that a local basis can be constructed as eigenfunctions of a differential (Sobolev) operator and make connection with covariance parametric functions of the form $\{C_\theta \mid \theta \in \Theta\}$. It is important to highlight that the choice of such operator, as detailed in Sect. 5.3.1, permits to control the intrinsic properties of U [19, 21]. Moreover, the popularity of GP stems primarily from two essential properties. First, a Gaussian process is completely determined by its mean and covariance functions. This property facilitates model fitting as only the first- and second-order moments of the process require specification. Second, solving the prediction problem is straightforward since the optimal predictor at an unobserved position/time is a linear function of the observed values. Without loss of generality, we briefly present some definitions and properties of the eigenfunctions and propose a method to compute optimal parameters [24].

The rest of the paper is organized as follows. In Sect. 5.2 we review the representation of curves and we present some necessary geometrical tools for shape analysis including the construction of geodesics vector field between two corresponding curves. In Sect. 5.3 we formulate the problem of extending the deformation field to the whole domain as searching for eigenfunction of a Sobolev operator $\mathcal{L}$. Then, we propose to search for the optimal parameters of $\mathcal{L}$ using a Gaussian process. Experiment results and the conclusion will be given in the last two sections.

5.2   Notations and Background

Let I be the unit interval and let $\mathbb{L}^2(I, \mathbb{R}^2)$ be the set of square integrable functions from I to $\mathbb{R}^2$. Let $\eta : I = [0, 1] \longrightarrow \mathbb{R}^2$ denotes a parametrized curve in $\mathbb{L}^2(I, \mathbb{R}^2)$ satisfying: (1) η is absolutely continuous and (2) $\dot{\eta} \in \mathbb{L}^2(I, \mathbb{R}^2)$. Note that absolute continuity is equivalent to requiring that $\dot{\eta}(t)$ exists for almost $t \in I$, that $\dot{\eta}(t)$ is summable, and that $\eta(t) = \int_0^t \dot{\eta}(s)ds$. In this section we are interested in studying the geometry of shapes using the q-function representation of continuous curves in $\mathbb{R}^2$. It is shown that it is an efficient representation for analyzing shapes of curves [3]. Furthermore, it is the representation in which the elastic metric reduces to a simple L^2 metric and the space of unit length curves becomes the unit Hilbert sphere.

Now, we shall take a finite set of parametrized curves $\eta_0, \ldots, \eta_n$ satisfying all conditions and consider the shape representation of each η_i, $i = 0, \ldots, n$ by their q-functions $q_0, \ldots, q_n$. Let $\mathcal{M}$ denote the space of these shapes. Indeed, $\mathcal{M}$ is an infinite-dimensional manifold with a Riemannian structure on it as will be detailed next. The first goal is to compute the geodesic deformation vector field between any two parametrized curves on $\mathcal{M}$.

5.2.1   Curve Representation and Shape Space

Let $\eta : I = [0, 1] \longrightarrow \mathbb{R}^2$ denotes a parametrized curve in $\mathbb{L}^2(I, \mathbb{R}^2)$ satisfying all the conditions. For the purpose of studying its shape, we will represent it using its q-function $q : I \longrightarrow \mathbb{R}^2$ defined as

$$q(s) = \frac{\dot{\eta}(s)}{\sqrt{||\dot{\eta}(s)||}} \in \mathbb{R}^2. \tag{5.1}$$

Here $||.||$ denotes the Euclidean 2-norm in $\mathbb{R}^2$. This vector valued function q is the tangent vector normalized by the square-root of the instantaneous speed along the curve and is a local descriptor of the geometry of the curve. The original curve η can be reconstructed using $\eta(s) = \int_0^s ||q(t)|| \, q(t) \, dt$. The scale-invariant shape representation is given by normalizing the function q by its magnitude as $\frac{q}{\sqrt{\int_0^1 ||q(s)||^2 ds}}$. Therefore, it becomes an element of a unit sphere in the Hilbert manifold $\mathbb{L}^2(I, \mathbb{R}^2)$ that we will denote as $\mathcal{M}$. This is an infinite-dimensional Hilbert manifold and represents the shape space of all translation and scale-invariant elastic curves. More precisely, we can define the manifold $\mathcal{M}$ in the form

$$\mathcal{M} \equiv \left\{ q \in \mathbb{L}^2(I, \mathbb{R}^2) | \int_0^1 (q(s), q(s))_{\mathbb{R}^2} ds = 1 \right\}. \tag{5.2}$$

Here and subsequently, $(,)_{\mathbb{R}^2}$ stands for the standard Euclidean inner product in $\mathbb{R}^2$.

5.2.2 Geodesics in Shape Space, Exponential Map, and Parallel Transport

An important geometrical construct for the analysis of shapes is the definition of the tangent space. Since M is a Hilbert sphere in $\mathbb{L}^2(I, \mathbb{R}^2)$, at any curve $q \in M$, we define the tangent space and we denote $T_q M$. We equip the tangent space of M with a smoothly varying Riemannian metric that measures infinitesimal lengths on the shape space. This inner product is first defined generally on $\mathbb{L}^2$ and then induced on the tangent space of M. The metric defined on M has a nice physical interpretation in being an elastic metric. More precisely, let f and g be two tangent vectors in $T_q M$, the metric is defined as,

$$\langle f, g \rangle = \int_0^1 (f(s), g(s))_{\mathbb{R}^2} ds. \tag{5.3}$$

Another important step in our shape analysis is to compute geodesic paths between shapes with respect to the chosen metric. With respect to the q-function, M is represented as the Hilbert sphere in $\mathbb{L}^2(I, \mathbb{R}^2)$ and obviously lot is known about the geometry of a sphere, including geodesics and exponential map. Therefore, geodesics between any two points q_1 and q_2 (not antipodal to q_1) on M are great circles and it is expressed in terms of a tangent direction $f \in T_{q_1} M$ as,

$$\chi_t(q_1; f) = \cos(t\|f\|) q_1 + \sin(t\|f\|) \frac{f}{\|f\|}. \tag{5.4}$$

This equation gives the constant-speed parametrization of the geodesic passing through q_1 with velocity vector f at $t = 0$. As a result, the exponential map $\exp : T_{q_1} M \longrightarrow M$ is defined as

$$\exp_{q_1}(f) = q_2 = \cos(\|f\|) q_1 + \sin(\|f\|) \frac{f}{\|f\|}. \tag{5.5}$$

The length of the geodesic determines an elastic quantitative distance between two shapes q_1 and q_2 in M given by

$$d_Q(q_1, q_2) = \cos^{-1}(\langle q_1, q_2 \rangle). \tag{5.6}$$

From Eq. 5.4, the velocity vector along the geodesic path χ_t is obtained as $\dot{\chi}_t$. It is also noted that $\chi_0(q_1) = q_1$ and $\chi_1(q_1) = \exp_{q_1}(f) = q_2$. Conversely, given two shapes q_1 and q_2, the inverse exponential map (also known as the logarithmic map) allows the recovery of the tangent vector f between them, and is computed as follows:

$$\exp_{q_1}^{-1}(q_2) = f = \frac{\cos^{-1}\langle q_1, q_2 \rangle}{\sin(\cos^{-1}\langle q_1, q_2 \rangle)}(q_2 - \langle q_1, q_2 \rangle q_1). \tag{5.7}$$

For any two points q_1 and q_2 on $\mathcal{M}$, the map $\Gamma : T_{q_1}\mathcal{M} \longrightarrow T_{q_2}\mathcal{M}$ parallel transports a vector f from q_1 to q_2 and is given by

$$\Gamma_{q_1 \longrightarrow q_2}(f) = f - 2\frac{(q_1 + q_2)\int_0^1 (f, q_2)ds}{\int_0^1 (q_1 + q_2, q_1 + q_2)ds}. \tag{5.8}$$

To summarize, the exponential map takes points in the tangent plane to points on the sphere, preserving distance from q; it also preserves the tangential direction from q. Concretely, the exponential map only preserves angles and distances for points in the tangent plane which have distance $\lesssim \pi$ from q; however, we shall implicitly assume this condition holds whenever it is needed. Given the above tools for constructing geodesics and inverse exponential maps on the shape space, we will indicate in the next section how these equations may be used to solve the problem of estimating a deformation vector field outside the landmark curves.

5.3 The Geometric Constraints on the Embedded Manifold

Let Ω be the bounded domain on $\mathbb{R}^2$ (usually $[0, 1]^2$), $\{\beta_j, \ j = 1, \ldots, N_l\}$ a finite set of N_l landmark curves on the target image I_2, and $\{\alpha_j, \ j = 1, \ldots, N_l\}$ a set of corresponding curves on the reference image I_1. Let $\Psi \in \mathcal{W}(\Omega, \mathbb{R}^2)$ be the required deformation vector field representing the registration and $\mathcal{W}$ a Sobolev space of sufficiently smooth vector fields over Ω with appropriate boundary conditions. We note that this method is easily generalized to $\mathbb{R}^n$, but the current application only warrants registration of 2D and 3D images. Using properties from the previous section, we note the restriction of Ψ on β_j as the geodesic deformation field bringing β_j to α_j in the corresponding shape space.

We begin by introducing a compact linear operator to ensure that the registration outside β_js is constructed by Gaussian process regression. The goal is to find a locally smooth deformation ϕ that maps any small neighborhood in Ω uniformly [17]. Thus, the registration problem can be formulated as follows: find an optimal deformation vector field $\Psi : \Omega \mapsto \Omega$ within a suitable function space $\mathcal{W}$ by minimizing the following functional:

$$E : \mathcal{W} \to \mathbb{R}^+$$

$$\Psi \mapsto E(\Psi) = \frac{\lambda}{2}\|\Psi\|_\mathcal{W} + \frac{1-\lambda}{2N_l}\sum_{j=1}^{N_l} d^2(\alpha_j, \Psi(\beta_j)), \tag{5.9}$$

where $\lambda \in [0, 1]$ and d is a geodesic distance between α_j and $\Psi(\beta_j)$. We use an elastic Riemannian framework to compute the geodesics between corresponding curves. The main advantages of this choice are that it is invariant to re-parametrizations of curves and the resulting geodesic distance has an intuitive interpretation in terms of the amount of stretching and bending needed to deform one curve into another. We refer the reader to [18] for more details.

5.3.1   Local Solutions as Eigenfunctions of the Covariance Operator

Here, for simplicity, we restrict our attention to the case of one landmark curve per image. In practice, each landmark curve is represented using a collection of points $\{p_i^1, \; i = 1, \ldots, N\} \in \alpha$ on I_1 and points $\{p_i^2, \; i = 1, \ldots, N\} \in \beta$ on I_2. Let U denote the shooting (geodesic) vector field taking curve β to curve α. Then, U_i is the displacement vector bringing point p_i^2 to point p_i^1: $U_i(p_i^2) = p_i^1$. Let $\mathcal{L}$ be a linear operator such that $\mathcal{L} = \mathcal{P}^*\mathcal{P}$, where $\mathcal{P}$ is a differential operator with adjoint $\mathcal{P}^*$. Then, the first term of Eq. 5.9 can be written as $\|\Psi\|_W = \int_\Omega \|\mathcal{P}\Psi\|$ and the solution of minimizing Eq. 5.9 such that $\nabla E(\Psi) = 0$ is given by

$$\begin{cases} \Psi = \sum_{i=1}^N b_i \phi_i & , b_i \in \mathbb{R} \\ \Psi(p_i^2) = p_i^1 \\ \mathcal{L}_i \phi_j = \delta_{ij}; \mathcal{L}_i = (-\Delta + \epsilon_i I)^m, & m \geq 1 \end{cases} \tag{5.10}$$

To maintain the constraint $\sum_{i=1}^N b_i^2 = 1$ we use a truncated version of the Karhunen–Loève expansion. We will briefly recall some related notions of the K-L expansion to make connection with the proposed formulation. Let $(\phi_j)_j$ be a set of orthonormal functions in $\mathbb{L}^2([0, 1])$, i.e.,

$$\langle \phi_k, \phi_j \rangle_{\mathbb{L}^2} = \int_0^1 \phi_k(t)\phi_j(t)dt = \delta_{kj} \tag{5.11}$$

where δ_{kj} denotes the Kronecker index function satisfying $\delta_{kj} = \begin{cases} 1, & \text{if } k = j \\ 0, & \text{if } k \neq j \end{cases}$.

Furthermore, the eigenvalues λ_j and the eigenfunctions ϕ_j of the operator $\mathcal{L}$ can be obtained by solving the Fredholm integral

$$(\mathcal{L}\phi_j)(t) = \lambda_j \phi_j(t) \tag{5.12}$$

To show examples of the link between Hilbert–Schmidt and Sturm–Liouville eigenvalue problems we give a list of specific examples. The choice of a covariance operator as well as the corresponding eigen-expansion has a key role in constructing local solutions. Illustrative examples are given in Fig. 5.1.

- **Case 1, $\epsilon_i = 0$:** $\mathcal{P} = \partial_x$, $\mathcal{L}$ reduces to a Laplacian operator, and the solution is given by thin plate splines (TPS). Note that this makes the proposed method more general than TPS [19].

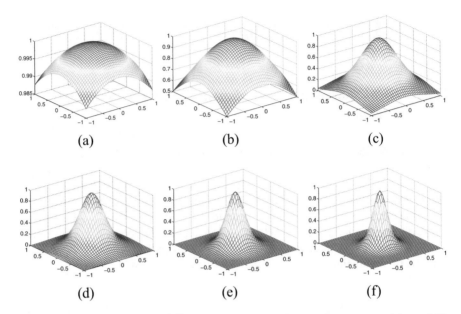

Fig. 5.1 Illustrative examples in $\mathbb{R}^3$ with different values of the local parameter ϵ. (**a**) $\epsilon = 0.25$, (**b**) $\epsilon = 2$, (**c**) $\epsilon = 5$, (**d**) $\epsilon = 8$, (**e**) $\epsilon = 11$, (**f**) $\epsilon = 14$

- **Case 2,** $m = 1$: $\mathcal{L} = (\epsilon_i^2 I - \frac{d^2}{dx^2})$ and the local solution is given by $\phi_i(x) = \frac{1}{2\epsilon_i^2} \exp(-\epsilon_i x)$ where ϵ_i is a parameter that determines the shape of the covariance function ϕ_i [20].
- **Case 3,** $m = 2$: $\mathcal{L} = (\epsilon_i^2 I - \frac{d^2}{dx^2})^2$ and the local solution is given by $\phi_i(x) = \frac{1}{8\epsilon_i^3}(1 + \epsilon_i x) \exp(-\epsilon_i x)$.
- **General Case:** The solution with respect to $\mathcal{L}$ is given by

$$\phi_i(x) = \frac{2^{1-m-\frac{1}{2}}}{\pi^{\frac{1}{2}} \Gamma(m) \epsilon_i^{2m-1}} (\epsilon_i \|x\|)^{m-\frac{1}{2}} K_{\frac{1}{2}-m}(\epsilon_i \|x\|),$$

where K is the modified Bessel function. We note that in all cases, $N_\phi = \{\phi_i\}$ is the native space of ϕ_is endowed with a Riemannian metric. For simplicity, we use $U = [U_1, U_2, \ldots, U_N]^\top$ and $b = [b_1, b_2, \ldots, b_N]^\top$. Let $A_{ij} = \langle \phi_i, \phi_j \rangle$ be the covariance matrix with A positive definite determined by the choice of eigenfunctions ϕ_i.

It is now clear that the better solution depends on different parameters, but all of them are related to the choice of $\mathcal{L}$. In fact, controlling the parameters allows us to take into consideration additional constraints derived from the properties of N_ϕ, i.e., the convergence of the finite approximation and the stability of its numerical implementation. Especially, the solution reduces to an exponential for small ϵ_i and to a Gaussian for large ϵ_i. Therefore, the solution includes a large class (family) of functions making it very flexible, which presents a big advantage for the proposed method [16]. Next we will use a non-deterministic formulation to compute the optimal parameters that determine the eigenfunctions which is the same as choosing the Sobolev operator $\mathcal{L}$.

5.3.2 Global Solution as a Random Field

Assuming we are only interested in the optimal solution at certain points p_i^2, say positions $(x_1, x_2, \ldots, x_n)$ on Ω. If we go back to the embedding problem from the previous section and rewrite it in a natural probabilistic setting, the solution of Eq. 5.10 turns out to have important optimality properties. Like in the previous section, we assume that we want to find the response $U(x) = \Psi(x) - x$ of an unknown model function Ψ at a new point x of a set Ω, provided that we have a sample of input-response pairs $(x_j, U_j) = (x_j, \Psi(x_j))$ given by points on curves. But now we assume that the whole setting is non-deterministic, i.e., the response U_j at x_j is a realization of a Gaussian process $U(x_j)$.

The short description in this section is limited to necessary notions that enable statistical inference on partial realization of random field, as it is usually the case in features-based registration as well as prediction problems. Indeed, spatial data contain information about both the attribute of interest as well as its location. The location may be a set of coordinates, such as the latitude and longitude, or it may be a small region such as curves.

Given a parameter space Ω, a random field U over Ω is a collection of random variables $\{U(X), X \in \Omega\}$. Since multivariate distributions are determined by their means and covariances, it is straightforward that random fields are determined by their mean and covariance functions, defined by

$$\mu(X) = \mathbb{E}\{U(X)\}$$

and

$$C(X, S) = \mathbb{E}\{(U(X) - \mu(X))^T (U(S) - \mu(S))\}$$

so that the elements of C are given by

$$C_{i,j}(X, S) = \mathbb{E}\{(U_i(X) - \mu_i(X))^T (U_j(S) - \mu_j(S))\}.$$

5.3.3   Numerical Solution

Let U be a Gaussian random field (as defined in the previous paragraph), i.e., for $X \in \Omega$:

$$U(X) \hookrightarrow \mathcal{N}(\mu_X, C_X) = \mu_X + \mathcal{N}(0, C_X),$$

where μ_X and C_X are the mean and the covariance function of U, respectively. When μ_X is often determined by the given conditions (here $U_i(p_i^2) = p_i^1$), C_X must be estimated as has been detailed in the previous section. There is a large choice of covariance functions in the literature, see [22]. Note that an important consequence of this is that the mathematical properties and the quality of the estimated field will change accordingly. In this work, we choose C to be the family of Matérn covariance functions (see the general solution of Eq. 5.9 for more details), i.e.:

$$C_X(h) = \sigma^2 \frac{2^{1-\nu}}{\Gamma(\nu)} (||h||/l)^\nu K_\nu(||h||/l) \tag{5.13}$$

which is the spatial correlation at distance $||h||$, where K is the modified Bessel function of the second kind. Generally, $\sigma^2 > 0$ is referred to as a (marginal) variance parameter, $l > 0$ as a (spatial) scale parameter, and $\nu > 0$ as a smoothness parameter. If $\nu = \frac{1}{2} + k, k \in \mathbb{N}$, Eq. 5.13 reduces to the product of an exponential function and a polynomial [23]:

$$C_X(h) = \sigma^2 e^{-||h||/l} \sum_{j=0}^{k} \frac{(k+j)!}{(2k)!} \binom{k}{j} (2||h||/l)^{k-j}$$

To make connection with the eigenfunctions of $\mathcal{L}$ operator in Eq. 5.9, we remind that for $\nu = \frac{1}{2}$, C becomes an exponential, and for $\nu = +\infty$ it becomes a Gaussian. In this work, parameters σ and l are determined using a maximum likelihood estimator (MLE): minimizing the negative log likelihood function [19, 24]:

$$-ln(L(X|l, \sigma^2)) = \frac{n}{2}ln(2\pi) + \frac{n}{2}ln(\sigma^2) + \frac{ln|V_{X,l}|}{2} + \frac{X^T V_{X,l}^{-1} X}{2\sigma^2}, \tag{5.14}$$

where

$$\sigma^2 V_{X,l}(h) = C_X(h).$$

To find the minimizer of Eq. 5.14 we use a Newton-based method to determine $\hat{\sigma}$ and $\hat{l}$ as well as a cross-validation technique to choose $\hat{\nu}$. Once all parameters are evaluated, $\hat{U}$ at an unobserved position is computed with the help of the circulant embedding method [25].

5.4   Numerical Examples

In the medical context, ultrasound (US) and magnetic resonance imaging (MRI) techniques are noninvasive diagnostic test that take detailed images of the soft tissues of the body in order to highlight area of interest. In the following example, the goal is to register images of different modalities (intensity distributions) in order to be able to fuse complementary information. Since no correspondence is available, the only solution is to exploit boundaries of apparent organs and use them as landmark curves. Thus, given US and MRI images, practitioners select a set of corresponding curves that define organ boundaries in both images. Without additional information, we use these curves for registering data. In the case of one landmark curve $M = 1$, we first register the corresponding two curves and then discretize them for a numerical solution. In the case of a set of curves, we apply the same idea to each pair of corresponding curves. First, we assume that the curve β_1 is sampled with a finite set of points $X = \{X_i, i = 1, \ldots, N = 100\}$ with corresponding deformations U_i; these deformations are estimated using the elastic geodesic path from β_1 to α_1 as detailed in Sect. 5.2. The goal is to estimate $\hat{U}$ outside the given landmark curves (i.e., for the entire bounded domain Ω) using the methodology of random fields. An example is shown in Fig. 5.2 top row where β_1 represents the curve in US (a) and α_1 represents the reference curve on MRI (b). In this example, we first compute the geodesic deformation vector field U such that

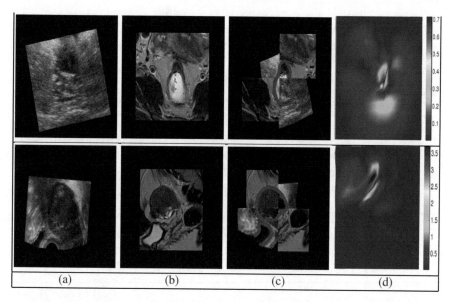

Fig. 5.2 Results of the proposed method applied to two examples. (**a**) Original US image before deformation, (**b**) MRI slice, (**c**) overlap of MRI image and the registered US, and (**d**) the Laplacian of the deformation vector field on the whole image domain (embedding)

$U(\beta_1) = \alpha_1$ on the curve domain $I = [0, 1]$, and then use the proposed method to compute Ψ or $(U = \Psi - id)$ on $\Omega \setminus I$. The same idea is applied in the case of two corresponding curves $M = 2$ in Fig. 5.2 bottom. In both examples, we show the Laplacian of the deformation to check the smoothness of the solution.

For a more general idea when using eigenfunctions of a linear operator and for comparative solutions, we use five different methods to perform the registration: the proposed method (**M**), thin plate splines (**TPS**), multi-quadratic (**MQ**), inverse multi-quadratic (**IMQ**), and Gaussian (**G**). We compute the initial displacement U by first finding the optimal correspondence (re-parametrization) between continuous landmark curves, and then the geodesic deformation vector field. Furthermore, for each method, we computed the optimal parameters. For fairness of comparisons, we provided the optimal parameters and the optimal correspondences between landmark curves for all methods (which improved their performance). All results are summarized in Table 5.1.

For illustration, Fig. 5.3 shows the original US and MRI images, the resulting deformed US image, the deformed grid, the deformation vector field and its corresponding Laplacian map, and two views of the MRI–US overlap after registration. First, the proposed registration process provides more precision (MRI–US overlap or fusion in Fig. 5.3f and g) about the locations around structures of interest, which consequently simplifies and improves the clinical diagnosis. As can be seen, the proposed method almost uniformly outperforms all of the other methods confirming that using the linear operator $\mathcal{L}$ makes our solution more general than TPS, exponential, and Gaussian-based methods.

To better visualize the smoothness of the resulting deformation vector fields in (d), we show their Laplacian maps in panel (h). If the deformation is smooth, we expect the Laplacian to be constant with values close to 0. It is evident that

Table 5.1 Registration accuracy on data of ten patients for different methods: our method (**M**), thin plate splines (**TPS**), inverse multi-quadratic (**IMQ**), multi-quadratic (**MQ**), and Gaussian (**G**)

		M		TPS		MQ		IMQ		G	
P	ν	RMSE	SE	RMSE	SE	RMSE	SE	RMSE	SE	RMSE	SE
1	7/2	**0.288**	**0.044**	0.413	0.050	*0.346*	*0.048*	0.366	0.050	3.495	0.092
2	3/2	**0.076**	**0.021**	*0.077*	0.021	0.209	0.051	0.604	0.298	6.046	*0.021*
3	3/2	**0.015**	**0.014**	*0.017*	*0.014*	0.021	0.015	0.763	0.046	0.824	0.046
4	5/2	**0.027**	**0.013**	0.042	0.020	*0.028*	*0.017*	0.044	0.021	0.0560	0.023
5	5/2	**0.066**	**0.017**	0.068	0.020	0.091	0.025	*0.066*	*0.020*	0.099	0.019
6	7/2	**0.025**	**0.020**	0.036	0.021	0.360	0.110	*0.028*	*0.020*	0.037	0.025
7	5/2	**0.093**	**0.030**	*0.099*	*0.030*	0.110	0.035	0.102	0.033	0.155	0.043
8	11/2	**0.033**	**0.011**	0.062	0.017	0.067	0.021	*0.048*	0.016	0.066	*0.015*
9	3/2	**0.069**	0.019	0.073	**0.019**	0.093	0.029	*0.071*	*0.020*	0.090	0.023
10	5/2	**0.029**	*0.020*	*0.033*	**0.020**	0.110	0.063	0.035	0.024	0.048	0.027

The best performance is highlighted in bold and the second best is italicized. The performance is given by root mean square error (**RMSE**) and shape error (**SE**) coefficients between the reference (MRI) and target (US) curves

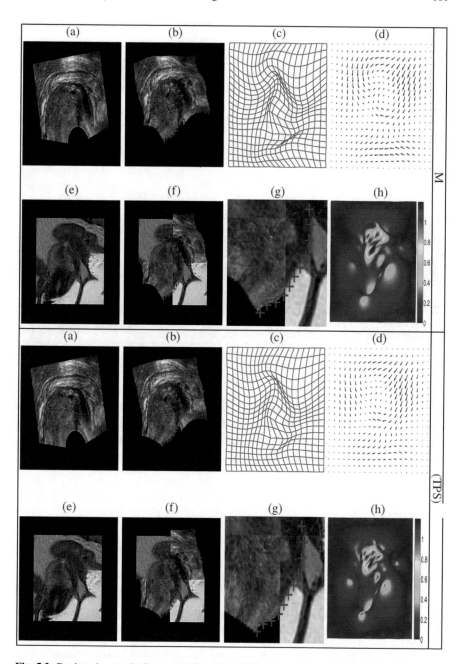

Fig. 5.3 Registration results for same patient computed using the proposed method (M), thin plate splines (TPS), and inverse multi-quadratic (IMQ). For each method, we show (**a**) the US image, (**b**) the deformed US image, (**c**) the deformed grid, (**d**) the deformation vector field, (**e**) the MRI image, (**f**) the MRI–US overlap, (**g**) a zoom-in of (**f**) for improved visualization, and (**h**) the Laplacian of the deformation vector field

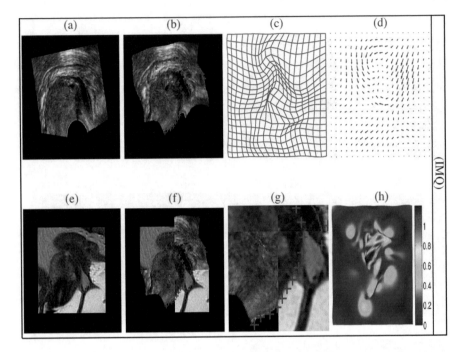

Fig. 5.3 (continued)

overall the proposed method generates smoother deformations with a higher level of accuracy than the other two methods. Even with small values, we expect that a part of the errors that occurred are due to numerical approximations on grids, especially when data points are very close.

References

1. Kendall, D. G. (1984) Shape manifolds, Procrustean metrics and complex projective spaces. Bulletin of the London Mathematical Society, 16, 81–121.
2. Klassen, E and Srivastava, A and Mio, W. and Joshi, S.H. (2004) Analysis of planar shapes using geodesic paths on shape spaces. IEEE Transactions on Pattern Analysis and Machine Intelligence, 26:2, 372–383.
3. A. Srivastava and E. Klassen and S. H. Joshi and I. H. Jermyn (2011) Shape Analysis of Elastic Curves in Euclidean Spaces. IEEE Transactions on Pattern Analysis and Machine Intelligence (PAMI), 33:7, 1415–1428.
4. Joshi, S.H. and Klassen, E. and Srivastava, A. and Jermyn, I. (2007) A novel representation for Riemannian analysis of elastic curves in $\mathbb{R}^n$. IEEE Computer Vision and Pattern Recognition (CVPR), Conference on, 1–7.

5. Joshi, S.H. and Klassen, E. and Srivastava, A. and Jermyn, I. (2007) Removing shape-preserving transformations in square-root elastic (SRE) framework for shape analysis of curves. EMMCVPR, 387–398.
6. Anuj Srivastava and Eric Klassen (2016) Functional and Shape Data Analysis. Springer, New York, NY.
7. P. W. Michor and D. Mumford (2006) Riemannian geometries on spaces of plane curves. Journal of the European Mathematical Society, 8, 1–48.
8. E. Sharon and D. Mumford (2006) $2D$-shape analysis using conformal mapping. International Journal on Computer Vision, 70:1, 55–75.
9. M.I Miller and L. Younes (2002) Group actions, homeomorphisms, and matching: A general framework. International Journal of Computer Vision, 41:1/2, 61–84.
10. A. Sotiras and C. Davatzikos and N. Paragios (2013) Deformable Medical Image Registration: A survey. IEEE Transactions on Medical Imaging, 32:7, 1153–1190.
11. S. Ying and Y. Wang and Z. Wen and Y. Lin (2016) Nonlinear $2D$ shape registration via thin-plate spline and Lie group representation. Neurocomputing, 195, 129–136.
12. D. Zosso and X. Bresson and J.-P. Thiran (2011) Geodesic active fields–a geometric framework for image registration. IEEE Transactions on Image Processing, 20:5, 1300–1312.
13. U. Grenander and M. I. Miller (1998) Computational anatomy: An emerging discipline. Quarterly of Applied Mathematics, LVI:4, 617–694.
14. M. Deshmukh and U. Bhosle (2011) A survey of image registration. IJIP, 5, 245–269.
15. Fred L. Bookstein (1989) Principal warps: thin-plate splines and the decomposition of deformations. Trans. PAMI 11:6, 567–585.
16. Bertil Matérn (1960) Spatial Variations. PhD Thesis, Stockholm University.
17. Grenander Ulf and Miller Michael (2007) Pattern Theory: From Representation to Inference. Oxford University Press, Inc.
18. A. Srivastava and E. Klassen and S.H. Joshi and I.H. Jermyn (2011) Shape analysis of elastic curves in Euclidean spaces. IEEE Trans. on PAMI, 33:7, 1415–1428.
19. M. Bozzini and M. Rossini and R. Schaback (2013) Generalized Whittle-Matérn and polyharmonic kernels. Advances in Computational Mathematics, 39:1, 129–141.
20. Tilmann Gneiting (2002) Compactly supported correlation functions. J. of Multivariate Analysis, 83, 493–508.
21. Stein Michael L. (1999) Interpolation of Spatial Data. Springer Series in Statistics.
22. Adler R. J. and Taylor Jonathan E. (2007) Random Fields and Geometry. Springer Monographs in Mathematics.
23. T. Gneiting and W. Kleiber and M. Schlather (2009) Matérn Cross-Covariance Functions for Multivariate Random Fields. Department of Statistics, University of Washington, 549.
24. P.K. Kitanidis and R.W. Lane (1985) Maximum likelihood parameter estimation of hydrologic spacial processes by the Gauss-Newton method. Journal of Hydrology, 79, 53–71.
25. C.R. Dietrich and G.N. Newsam (2006) A fast and exact method for multidimensional Gaussian stochastic simulations: Extension to realizations conditioned on direct and indirect measurements. Water Resources Research, 32:6, 1643–1652.

Chapter 6
Numerical Study of SIF for a Crack in P265GH Steel by XFEM

Houda Salmi, Khalid El Had, Hanan El Bhilat, and Abdelilah Hachim

Abstract The analytical solving of fracture mechanics equations remains insufficient for complex mechanisms, hence the use of finite element numerical methods (FEM). But the presence of singularities strongly degrades the FEM convergence and refining the mesh near the singularities is not enough to obtain an accurate solution, hence the use of the extended finite element method (XFEM). With XFEM, the standard finite element approximation is locally enriched by enrichment functions to model the crack. The present work focuses on the numerical study of the defects harmfulness in the P265GH steel of a Compact Tension (CT) specimen. A stress intensity factor (SIF) was calculated by CAST3M code, using XFEM and the G-Theta method in the FEM; the objective is to simulate a CT sample with XFEM in 3D and to calculate the critical length of crack leading to the fracture as well as the evolution of stress concentration coefficient. An integration strategy and a definition of level sets have been proposed for cracks simulation in XFEM. A weak loading was considered to ensure elastic behavior. A comparative study of the numerical SIF values with the theory was performed. The result shows that XFEM is a precise tool for modeling crack propagation.

Keywords Fracture · Stress intensity factor · XFEM · CAST3M code · G-Theta method · P265GH steel

H. Salmi (✉) · H. El Bhilat
National Higher School of Electricity and Mechanics, Laboratory of Control and Mechanical Characterization of Materials and Structures, Casablanca, Morocco

K. El Had · A. Hachim
Higher Institute of Maritime Studies, Laboratory of Mechanics, Casablanca, Morocco

© Springer Nature Switzerland AG 2020
S. Dos Santos et al. (eds.), *Recent Advances in Mathematics and Technology*,
Applied and Numerical Harmonic Analysis,
https://doi.org/10.1007/978-3-030-35202-8_6

6.1 Introduction

The analytical solving of the equations remains limited to simple problems whereas in reality there are complex mechanisms hence the use of finite element method (FEM). This method consists of splitting the spatial domain into smaller, simpler parts (finite elements) and looking for a simplified formulation of the problem on each element. The simple equations that model these finite elements are then assembled into a larger system of equations that models the entire problem [1]. Many studies suggest that crack propagation occurs over approximately 90% of the life of a component [2, 3]. Therefore, it is necessary to be able to evaluate the evolution of defects during the loading as well as their critical sizes, this knowledge will allow the establishment of adequate inspection and maintenance programs. The problem of defects harmfulness is approached by global parameters such as the stress intensity factor (SIF) in elasticity or the J integral in plasticity [4, 5]. The Linear Elastic Fracture Mechanics (LEFM) [5, 6] is the basic theory of fracture for characterizing the stress fields near the crack tip. The analysis of the materials fracture remains one of the most difficult problems for numerical computational mechanical methods, the need for adequate design of high performance and reliable structural components highlights the importance of accurate modeling of crack initiation and propagation in numerical fracture analysis. The finite element method (FEM), as an effective numerical method in computational mechanics, remains limited in terms of modeling crack growth, mainly due to the incremental re-meshing of the crack and other convergence considerations. Over the last few decades, several methods have been introduced to replace the FEM, hence the development of the Extended Finite Element Method (XFEM), which was inspired by the partition of the unit finite element method (PUFEM) [7]. In this method, the finite element basis is enriched by functions that describe the separation of the material and the singularity, the elements cut by the crack are presented by special elements with additional degrees of freedom. In XFEM the geometry of the crack can be located arbitrary through the mesh, and crack growth simulations can be performed without any required re-meshing [8]. XFEM was originally introduced by Belytschko and Black [9] for modeling elastic crack growth. Moes et al. [10] and Dolbow [11] have improved this method by constructing an enriched approximation that allows representing the crack independently of the mesh. The XFEM has presented advantages over the competing methods like the methods of elimination, re-meshing, etc. However, the XFEM presents certain difficulties. In fact, the presence of discontinuous functions in an element requires a specific integration strategy to describe the associated fields correctly. The most widespread integration strategy has been proposed by Moes et al. [12], it consists of cutting the enriched elements into sub-triangles and then applying a standard integration scheme to these triangles. Samaniego et al. [13] used this method in the case of a material with nonlinear behavior for shear bands modeling but this method does not ensure the conservation of energy around the crack tip. In order to maintain this energy, Elguedj [14] and Benoit Prabel [15] proposed in their work to use only quadrature elements during the calculation. For an effective advance of the crack, Prabel [15] defined also the level set on a finer intermediate grid not connected to the mesh.

The purpose of this study is to examine, in the elastic case, the defects harmfulness in the P265GH steel (this type of steel mainly used for pressurized equipment). In three-dimensional case, a numerical study using the FEM and the XFEM solver CAST3M [16] was performed considering tensile stress. The numerical method XFEM was illustrated by mathematical analysis using matrix form.

We modified the level set definition and the integration strategy to facilitate the use of the XFEM within CAST3M. The propagation is managed in a simple way by extending the mesh of the crack. We carried out a low load to ensure elastic behavior , we then compared the numerical stress factor intensity values with the theory and determined the critical length of the crack leading to the fracture. We finally presented the evolution of the stress concentration coefficient according to the axis of the specimen in 3D.

6.2   FEM Methodology

In numerical analysis, the finite element method (FEM) is used to solve partial differential equations, CAST3M is a structure calculation code using the finite element method to solve equations with boundary conditions, giving as solution the displacement function u, for an imposed tension, the system to be solved is reduced to Eq. (6.1):

$$Ku = f,$$
$$Cu = q \quad \text{(boundary conditions)}, \tag{6.1}$$

- with K: stiffness matrix,
- q: Vector column of imposed forces (nodal force),
- f: Vector of generalized forces,
- C: Elastic model tensor.

In FEM the displacement approximation u can be written as:

$$u(x) = \sum_{i \in N} N_i(x) u_i, \tag{6.2}$$

with

- N: set of all nodes in the domain,
- $N_i(x)$: standard finite element shape functions of node i,
- u_i: unknown of the standard finite element part at node i.

6.2.1 Local Approach: Stress Intensity Factor

The stress intensity factor characterizes the stress fields at crack tip. Stress in the vicinity of a crack admits an asymptotic development in the form [17]:

$$\sigma_r = \frac{K_I}{4\sqrt{2\pi r}}[5\cos(\frac{\theta}{2}) - \cos(\frac{3\theta}{2})] + \frac{K_{II}}{4\sqrt{2\pi r}}[5\sin(\frac{\theta}{2}) + 3\sin(\frac{3\theta}{2})] + o\sqrt{r}$$

$$\sigma_\theta = \frac{K_I}{4\sqrt{2\pi r}}[3\cos(\frac{\theta}{2}) + \cos(\frac{3\theta}{2})] - \frac{K_{II}}{4\sqrt{2\pi r}}[3\sin(\frac{\theta}{2}) + 3\sin(\frac{3\theta}{2})] + o\sqrt{r}$$

(6.3)

$$\sigma_{r\theta} = \frac{K_I}{4\sqrt{2\pi r}}[\sin(\frac{\theta}{2}) + \sin(\frac{3\theta}{2})] - \frac{K_{II}}{4\sqrt{2\pi r}}[\cos(\frac{\theta}{2}) + \cos(\frac{3\theta}{2})] + o\sqrt{r}.$$

With (r, θ) are polar coordinates of a point in the vicinity of the crack tip (Fig. 6.1).

The stress intensity factor is, therefore, expressed as a function of the stress field in the vicinity of the crack and its geometry, this stress field is numerically accessible by finite element, but (FEM) alone does not allow a precise approximation. We must, therefore, pay particular attention to the crack tip by using a particular finite element, type Barsoum [18], to force the singularity, Barsoum [18] proposes a finite element by displacing the nodes of the medium corresponding to the tip of the crack (edges 1–2 and 1–4) to a quarter of the length (Fig. 6.2).

6.2.2 Global Approach G-Theta Method (G_θ)

Energy release rate (G) represents the energy required to advance the crack at a unit length. It corresponds to the decrease of the total potential energy P_e to go from an

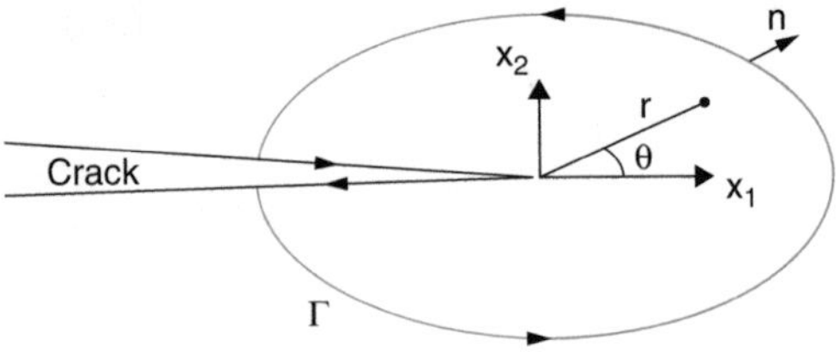

Fig. 6.1 Crack

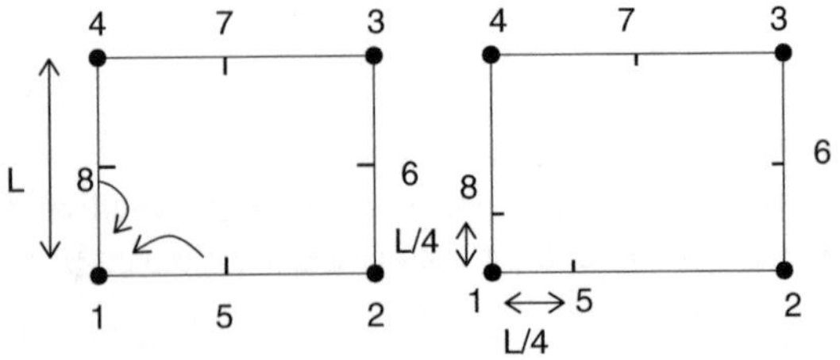

Fig. 6.2 Displacement of midpoints to quarter of the length

initial configuration with a length of crack a, to another where the crack has spread a length Δa:

$$\lim \frac{\Delta P_e}{\Delta a} = \frac{\partial P_e}{\partial a}, \tag{6.4}$$

with $P_e = F - U$ represents the potential energy.

- F: the work of external forces,
- U: the energy of elastic deformation of the solid [17].

The numerical calculation of Energy Release Rate consists in putting the expression of the Eq. (6.4) in the form of a contour integral.

Let Γ be a contour encircling the crack tip, if the lips of the crack are free of loading we have

$$J = G = \int_{\Gamma} \left(\frac{1}{2} Tr[\sigma \epsilon(u)]n_1 - \sigma n \frac{\partial u}{\partial x_1} \right) dS, \tag{6.5}$$

with

- u: displacement vector in a point M of the contour,
- n_1 and x_1: The coordinates of the point M of the integration contour relative of the crack front.
- dS: Contour element,
- σ: stress,
- ϵ: strain.
- n: normal vector in the point M.

6.2.3   Numerical Calculation of the Stress Intensity Factor (K)

Energy release rate (G) is related to the stress intensity factor of the mode I, in plane strain by Eq. (6.6):

$$G = \frac{(1 - v^2)K^2}{E}, \tag{6.6}$$

where E is the Young's modulus and v is the Poisson's ratio.

6.2.4   Geometry

The study considered elastic behavior of the material in P265GH steel, this steel is especially used in pressure equipment with the following properties:

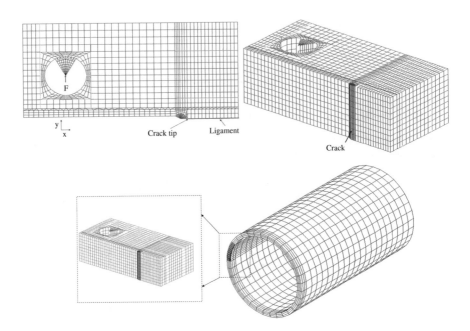

Fig. 6.3 Loading and boundary conditions

$$E \text{(Young modulus)} = 200.000 \text{ MPa},$$

$$\upsilon \text{(Poisson's ratio)} = 0.3,$$

$$\sigma_n \text{(nominal stress)} = 148 \text{ MPa},$$

$$\sigma_e \text{(yield stress)} = 320 \text{ MPa},$$

$$\sigma_u \text{(Breaking stress)} = 470 \text{ Mpa}.$$

We consider a Compact Tension P265 GH steel sample containing a crack and extracted from cylindrical pressure equipment (Fig. 6.3), initial length of crack is $a_0 = 11$ mm ($\frac{a_0}{W} = 0.55$). The sample is subjected to a tensile stress . The average value of the fracture toughness is $K_c = 96$ Mpa$\sqrt{m}$ [21]. The dimensions of the compact tension sample as per Fig. 6.4 are: $B = 10$ mm, $W = 2 \times B$, $W_1 = 2.5 \times B$, $a = 0.5 \times W$, $H = 1.2 \times B$, $H_1 = 0.65 \times B$, $D = 0.5 \times B$.

The movements of the mesh of the red segment in Fig. 9.3 are blocked. We applied a load of 166 Mpa to ensure elastic deformation. The loading is applied on the CT sample by means of a pin in the form of a rigid triangle to avoid any bending or torsion and to ensure that the tensile force is perfectly axial.

Fig. 6.4 Dimensions of the
CT specimen

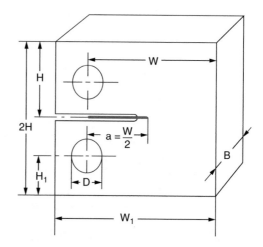

6.2.5   Meshing of the CT Specimen

The specimen has a plane of symmetry, so only half is modeled. In CAST3M, in 3D, to generate surfaces, we used the operator Dall. It is necessary that the opposite sides have the same number of elements. We use the operator Volu to generate the volume. Moreover to avoid the risk of tearing the meshes, we used the operator cout, which generates triangular elements of small surface. This surface makes it possible to weld the elements of the structure. For an exact approximation we used elements of Barsoum [18] at the crack tip. For a fine mesh, our model comprises eight slices ($N_t = 16$), this parameter does not influence the results too much, we took the radius of the circle capping the crack tip $Rc = 0.2$ mm, the model also contains $N_c = 5$ of concentric circles, the parameter N_c has a lot of influence on the results. The angle of the crack $\frac{\alpha}{2}$ is 60; the more $\frac{\alpha}{2}$ increases, the more the stress at the crack tip increases (Fig. 6.5).

We chose a mesh size of the crack tip of 0.15. In 3D the crack is a plane (Fig. 6.3) and the mesh is made up of 10,880 finite elements CUB8 containing eight Gauss points.

6.3   XFEM Methodology

6.3.1   Mathematical Formulation of XFEM

In XFEM, the standard finite element approximation is locally enriched to model the discontinuities. At a particular node of interest x_i, the displacement approximation u can be written as [10]:

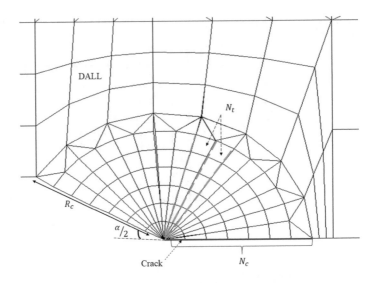

Fig. 6.5 Crack meshing

$$u(x) = \sum_{i \in N} N_i(x)u_i + \sum_{i \in N_d} N_i(x)(H(x) - (H(x_i))a_i$$

$$+ \sum_{i \in N_p} [N_i(x)(\sum_{\alpha=1}^{4} (\beta(x) - \beta(x_i))b_i^\alpha)], \tag{6.7}$$

with

$N_i(x)$: standard finite element shape functions of node i,
u_i: unknown of the standard finite element part at node i,
N: set of all nodes in the domain,
$N_d \subset$ N: Nodal subset of the Heaviside enrichment function $H(x)$ which is defined
for those elements entirely cut by the crack surface:

$$H(x) = -1, \quad \text{if} \quad \phi \le 0 \tag{6.8}$$
$$H(x) = 1, \quad \text{if} \quad if\phi > 0$$

Where $\phi(x)$ is the level set function.
a_i: unknown of the enrichment $H(x)$ at node i. These nodes are surrounded by a
square in Fig. 6.6.
$N_p \subset$ N: Nodal subset of the $\beta_\alpha(x)$ enrichment which are defined for those elements
partly cut by the crack front, four enrichment functions are used [19]:

Fig. 6.6 Description of the enrichment strategy

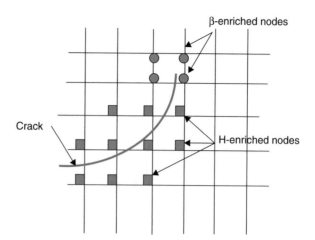

Fig. 6.7 Domain Ω with crack subjected to loads with all kinds of boundary at t time

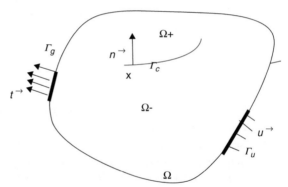

$$\beta_\alpha(r, \theta) = [\beta_1, \beta_2, \beta_3, \beta_4]$$

$$= [\sqrt{r}\sin(\frac{\theta}{2}), \sqrt{r}\cos(\frac{\theta}{2}), \sqrt{r}\sin(\frac{\theta}{2})\sin(\theta), \sqrt{r}\cos(\frac{\theta}{2})\cos(\theta)] \tag{6.9}$$

b_i: unknown of the enrichment $\beta_\alpha(x)$ at node i. These nodes are surrounded by a circle in Fig. 6.6.

We consider an elastic solid subjected to surface forces or displacements, as well as volume forces, we look to determine the stresses and strains at each point. These quantities are expressed by tensors that are written in matrix form. These are determined from the different relationships that link stresses, strains, and displacements. This approach leads to a system of partial differential equations that must be integrated taking into account the boundary conditions expressed in terms of force and/or displacements on the surface of the solid (Fig. 6.7).

The governing equations in static in the Cartesian coordinate system are

$$\nabla \sigma + b = 0 \quad or \quad (\sigma_{ij,j} + b_i = 0) \quad on \quad \Omega = \Omega^+ \cup \Omega^- \tag{6.10}$$

$$\sigma_{ij,j} n = t_i \quad on \quad \Gamma_g \tag{6.11}$$

$$U_i = u_i \quad on \quad \Gamma_u \tag{6.12}$$

$$\sigma_{ij,j} n = 0 \quad on \quad \Gamma_c \tag{6.13}$$

$$\varepsilon_{ij} = \frac{1}{2}(u_{i,j} + u_{j,i}) \tag{6.14}$$

$$\varepsilon_{ik,jl} + \varepsilon_{jl,ik} - \varepsilon_{il,jk} - \varepsilon_{jk,il} = 0 \tag{6.15}$$

$$\sigma_{ij,j} = c_{ij,kl}\varepsilon_{kl}, \tag{6.16}$$

where

- $\Omega \epsilon \mathbb{R}^3$ and n is a unit normal vector on the surface, b_i is the body force per unit volume, and Γ_g, Γ_u, and Γ_c are the traction, displacement, and crack boundaries, respectively.
- Equation (6.10) corresponds to the equilibrium equation where σ_{ij} is the Cauchy stress tensor at any point of the studied solid.
- Equations (6.11), (6.12) and (6.13) are obtained from the equilibrium on the surfaces (boundary conditions), t_i is the tensile force applied to the considered surface, u is the displacements field of displacements at any point of the solid, U is the displacement imposed on the point x of the surface.
- Equation (6.14) is a geometric equation that defines the deformations in the hypothesis of small perturbations, ε_{ij} is the strain tensor at any point of the solid.
- Equation (6.15) is the equation of compatibility so that we obtain six equations by circular permutation of indices (ijkl).
- Equation (6.16) is a constitutive law of linear elastic type where $c_{ij,kl}$ is tensor coefficient of the elastic constants.
- In numerical computation, we often look to minimize the elastic potential energy by defining the optimal stresses and strains.
- Multiplying the equation of equilibrium Eq. (6.10) by a test function δu and the integration on the domain of study Ω lead to:

$$\int_\Omega \nabla(\sigma.\delta u)d\Omega - \int_\Omega \sigma : \nabla(\delta u)d\Omega + \int_\Omega b\delta u d\Omega = 0. \tag{6.17}$$

Using the divergence theorem and the law of conservation of angular momentum:

$$\int_{\Gamma_g \cup \Gamma_u} t\delta u d\Gamma + \int_{\Gamma_c} t\delta u d\Gamma - \int_\Omega \sigma : (\delta\varepsilon)d\Omega + \int_\Omega b\delta u d\Omega = 0. \tag{6.18}$$

Because $\sigma_{ij,j}n = t_i$ on Γ_g denotes the prescribed traction, $\sigma_{ij,j}n = 0$ on Γ_c denotes the traction-free surface of the crack, and, knowing that test functions vanish on Γ_u, we have the weak form of the linear momentum equation for the continuum problem:

$$\int_{\Omega} \sigma : (\delta\varepsilon)d\Omega = \int_{\Gamma_g} t\delta u d\Gamma + \int_{\Omega} b\delta u d\Omega = R_{ext}. \tag{6.19}$$

Equation (6.19) can be further simplified as [20]:

$$\delta u^T K u = \int_{\Gamma_g} t\delta u d\Gamma + \int_{\Omega} b\delta u d\Omega \tag{6.20}$$

$$= R_{ext} \tag{6.21}$$

$$\text{With} \quad R_{ext} = \int_{\Gamma_g} t\delta u d\Gamma + \int_{\Omega} b\delta u d\Omega = \delta u^T f_{ext} \tag{6.22}$$

Then the equation to resolve by CAST3M is

$$K u = f_{ext}. \tag{6.23}$$

u is the displacement vector, f_{ext} is the external force vector, and K is the material stiffness matrix such that

$$K = \int_{\Omega} B^T D B d\Omega \tag{6.24}$$

$$f_{ext} = \int_{\Omega} b d\Omega + \int_{\Gamma_g} t d\Gamma. \tag{6.25}$$

The sub-matrices K_{ij} and f_i are obtained by substituting the approximation function defined in Eqs. (6.7), (6.24) and (6.25) [20]:

$$K_{ij} = \begin{bmatrix} K_{ij}^{uu} & K_{ij}^{ua} & K_{ij}^{ub} \\ K_{ij}^{au} & K_{ij}^{aa} & K_{ij}^{ab} \\ K_{ij}^{bu} & K_{ij}^{ba} & K_{ij}^{bb} \end{bmatrix} \quad and f_{iext}^h = [f_i^u, f_i^a, f_i^{b_1}, f_i^{b_2}, f_i^{b_3}, f_i^{b_4}]^T. \tag{6.26}$$

The sub-matrices and vectors that appear in the foregoing Eq. (6.26) are given as:

$$K_{ij}^{kl} = \int_{\Omega} (B_i^k)^T D(B_j^l) d\Omega \qquad where \quad k, l = a, u, b \tag{6.27}$$

$$f_i^u = \int_{\Omega} N_i b_i d\Omega + \int_{\Gamma_g} N_i t_i d\Gamma \tag{6.28}$$

$$f_i^a = \int_{\Omega} N_i (H(x) - H(x_i)) b_i d\Omega + \int_{\Gamma_g} N_i (H(x) - H(x_i)) t_i d\Gamma \tag{6.29}$$

$$f_i^{b\alpha} = \int_\Omega N_i(\beta_\alpha(x) - \beta_\alpha(x_i))b_i \, d\Omega + \int_{\Gamma_g} N_i(\beta_\alpha(x) - \beta_\alpha(x_i))t_i \, d\Gamma \quad (\alpha = 1, 2, 3, 4)$$
(6.30)

$$with \qquad B_i^u = \begin{bmatrix} N_{i,x} & 0 & 0 \\ 0 & N_{i,y} & 0 \\ 0 & 0 & N_{i,z} \\ 0 & N_{i,z} & N_{i,y} \\ N_{i,z} & 0 & N_{i,x} \\ N_{i,y} & N_{i,x} & 0 \end{bmatrix}$$
(6.31)

$$B_i^a = \begin{bmatrix} N_i(H(x) - H(x_i))_{,x} & 0 & 0 \\ 0 & N_i(H(x) - H(x_i))_{,y} & 0 \\ 0 & 0 & N_i(H(x) - H(x_i))_{,z} \\ 0 & N_i(H(x) - H(x_i))_{,z} & N_i(H(x) - H(x_i))_{,y} \\ N_i(H(x) - H(x_i))_{,z} & 0 & N_i(H(x) - H(x_i))_{,x} \\ N_i(H(x) - H(x_i))_{,y} & N_i(H(x) - H(x_i))_{,x} & 0 \end{bmatrix}$$
(6.32)

$$B_i^{\alpha b} = [B_i^{b_1}, B_i^{b_2}, B_i^{b_3}, B_i^{b_4}]$$
(6.33)

$$B_i^b = \begin{bmatrix} N_i(\beta_\alpha(x) - \beta_\alpha(x_i))_{,x} & 0 & 0 \\ 0 & N_i(\beta_\alpha(x) - \beta_\alpha(x_i))_{,y} & 0 \\ 0 & 0 & N_i(\beta_\alpha(x) - \beta_\alpha(x_i))_{,z} \\ 0 & N_i(\beta_\alpha(x) - \beta_\alpha(x_i))_{,z} & N_i(\beta_\alpha(x) - \beta_\alpha(x_i))_{,y} \\ N_i(\beta_\alpha(x) - \beta_\alpha(x_i))_{,z} & 0 & N_i(\beta_\alpha(x) - \beta_\alpha(x_i))_{,x} \\ N_i(\beta_\alpha(x) - \beta_\alpha(x_i)))_{,y} & N_i(\beta_\alpha(x) - \beta_\alpha(x_i))_{,x} & 0 \end{bmatrix}$$
(6.34)

$$D = \frac{E}{(1+\upsilon)(1-2\upsilon)} \begin{bmatrix} (1-\upsilon) & \upsilon & \upsilon & 0 & 0 & 0 \\ \upsilon & (1-\upsilon) & \upsilon & 0 & 0 & 0 \\ \upsilon & \upsilon & (1-\upsilon) & 0 & 0 & 0 \\ 0 & 0 & 0 & \frac{1-2\upsilon}{2} & 0 & 0 \\ 0 & 0 & 0 & 0 & \frac{1-2\upsilon}{2} & 0 \\ 0 & 0 & 0 & 0 & 0 & \frac{1-2\upsilon}{2} \end{bmatrix},$$
(6.35)

where

$N_i(x)$ are the standard finite element shape functions of node i.

B_i are the matrix of shape function derivatives, it is calculated at the Gauss points of each element.

$b_i(x)$ are the body force components.

$t_i(x)$ is the tensile force applied to surface.

Ω is the volume of the solid.

g is the area of the solid on which the tensile stress is applied.

Γ_g is the traction boundary.

$H(x)$ is the Heaviside function. $\beta_\alpha(x_i)$ is crack tip asymptotic enrichment functions.

D is the Hooke tensor in plane strain. K is the matrix of material rigidity.

E is the Young's modulus and υ is the Poisson's ratio.

6.3.2   Numerical Integration

The integration strategy is the use of triangular elements [12] (Fig. 6.8a) but in our integration strategy we looked for multiplying the Gauss points, so, we used quadrangle elements (Fig. 6.8b). The integration strategy in our numerical study is a simplified version of that proposed in the work of Prabel [15], which used quadrangle elements. This set of points is here introduced from the beginning of the calculation in CAST3M, in fact for a precise integration, the approach most used in most industrial calculation codes including CAST3M, is to split the initial element into several sub-elements each containing several Gauss points.

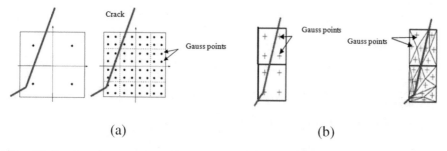

(a)   (b)

Fig. 6.8 Regular sub-division: (**a**) quadrangle elements; (**b**) Triangle elements [12]

Notes

For quadrangle elements, we use in XFEM (XQ4R) element, containing 4 Gauss points in two dimensions, and (XC8R) element, containing 8 Gauss points in three dimensions.

For triangular elements, we use (TRI3) containing 3 Gauss points in two dimensions, and (TRI6) element, containing 6 Gauss points in three dimensions.

6.3.3   Crack Meshing

In order to compare the XFEM with FEM we studied the same P265GH steel CT specimen with the same dimensions, the same mesh, the same loading, and boundary conditions. Except that in the XFEM modeling we have changed the type of element in the vicinity of the crack and we defined the level set function and the enrichment zone.

In order to minimize the simulation time, we used 280 quadrangle elements (XC8R) (orange, Fig. 6.9) from the crack tip to the end of the specimen. It should be noted that it is better not to extend the enrichment zone to the whole structure since it adds many additional degrees of freedom so the calculation may stop abruptly because of zero pivots. For the rest of the meshed domain, we used 10,800 standard elements CUB8 in the form of hexahedron with eight nodes. We choose a mesh size of crack tip of 0.15.

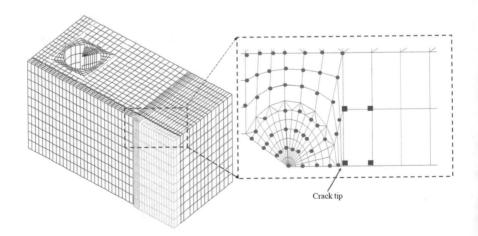

Fig. 6.9 Enrichment zone

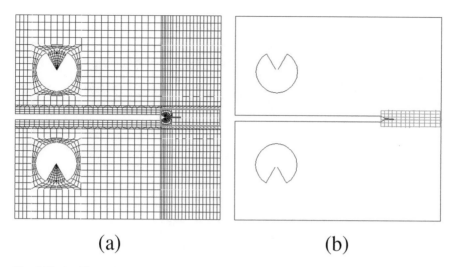

Fig. 6.10 Auxiliary grid method for calculating levels sets: (**a**) Mesh of the structure, (**b**) Auxiliary grid [15]

In order to compare between integration strategy using quadrangle elements and triangle elements, we use 280 triangle elements (TRI6) from the crack tip to the end of the specimen and 10,800 standard elements CUB8 for the rest of the mesh.

6.3.4   Enrichment and Level Sets

In order to effectively reproduce the advance of the crack Prabel [15] has defined the level set on a finer intermediate grid not connected to the mesh of the structure (Fig. 6.10).

We modified the definition of the level set by calculating them directly from the crack mesh without the need for an intermediate grid to reproduce the crack advance as proposed by Prabel [15] (Fig. 6.10), that is to say from a crack mesh, we define the normal level set from the crack front and the tangential level set from the crack lip (Figs. 6.11b and 6.12), such as the normal level set ϕ (PHI) gives the distance of a point x to the surface of the crack and the tangential level set ψ (PSI) gives the distance of a point x to the crack tip (Fig. 6.11), these level functions define the crack by (Fig. 6.13)

$$x \in \text{crack} \implies \phi(x) = 0 \quad \text{and} \quad \psi(x) \leq 0$$

$$\text{with} \quad (\mid \nabla\phi \mid = \mid \nabla\psi \mid = 1). \tag{6.36}$$

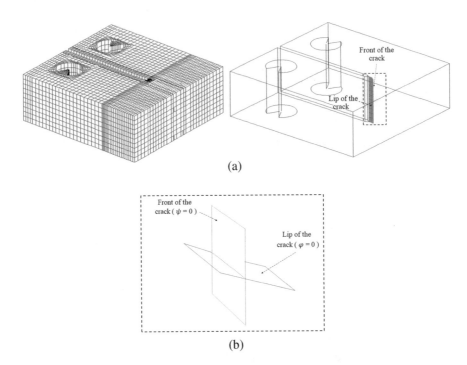

(a)

(b)

Fig. 6.11 Crack mesh: (**a**) Meshing of CT sample, (**b**) Definition of level set from crack mesh

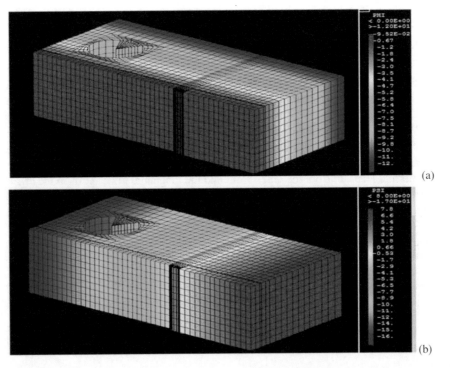

Fig. 6.12 Level sets in CASTEM (**a**) Normal level set ϕ (PHI), (**b**) Tangential level set ψ (PSI)

Fig. 6.13 Representation of
a crack with level sets

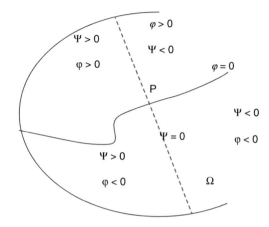

6.4   Results and Discussion

6.4.1   Calculation of Stress Intensity Factor (K)

The CAST3M 2017 software was used for modeling and simulation. It incorporates
G-Theta method for calculating G in elasticity along the crack front, for plane strain
we have

$$G = \frac{K^2}{E'} \quad and \quad E' = \frac{E}{(1 - v^2)}. \tag{6.37}$$

In order to validate the integration strategy and the modification of the level set
definition in XFEM, we calculated numerically the stress intensity factor (Fig. 6.14)
with the G-theta method using FEM and XFEM for a mesh of the crack by quadratic
and triangular elements. The results were compared to the analytical solution
presented in Eq. (6.38)[22].

$$K = \frac{F}{t\sqrt{w}}[29.6(\frac{a}{w})^{\frac{1}{2}} - 185.5(\frac{a}{w})^{\frac{3}{2}} + 655,7(\frac{a}{w})^{\frac{5}{2}} - 1017(\frac{a}{w})^{\frac{7}{2}} + 638,9(\frac{a}{w})^{\frac{9}{2}}],$$

$$\tag{6.38}$$

where F is the force applied in N, K is the stress intensity factor in $Mpa\sqrt{m}$, t is
the thickness of the CT sample and w its width in mm, a is the length of the crack
in mm.

To know the precision and the convergence of XFEM calculations in the CT
sample, we calculated the relative error e_K between K_{XFEM} calculated with XFEM
and K_a calculated with the analytical solution according the formula:

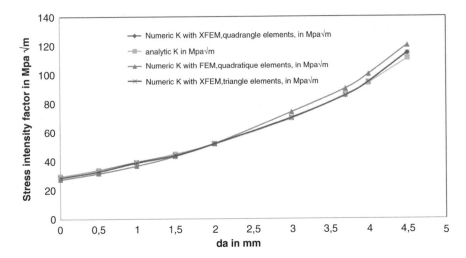

Fig. 6.14 Evolution of stress intensity factor versus the advanced crack

$$e_{K_1} = \frac{(K_{XFEM, Quad\ elemt} - K_a)}{K_a} \times 100$$

$$e_{K_2} = \frac{(K_{FEM, Quad\ elemt} - K_a)}{K_a} \times 100 \qquad (6.39)$$

$$e_{K_3} = \frac{(K_{XFEM, Triangle\ elemt} - K_a)}{K_a} \times 100.$$

Figure 6.15 shows that the absolute value of the relative difference between XFEM results and the corresponding values of the analytical solution are between 0 and 1% for quadrangle elements and between 0.5 and 3% for triangle elements, while the difference between the FEM results and analytical values is between 0.5 and 7%, which proves that our proposition of integration strategy and our modification of the level sets definition are valid and that its error is lower than 1%, so our modification in XFEM approaches the solution precisely.

Figure 6.16 presents the evolution of the stress intensity factor calculated by XFEM according to the progress of the crack for the applied stresses $\sigma = 166$ MPa and $\sigma = 183$ MPa.

Figure 6.16 shows that the stress intensity factor increases with the increase of the crack size and the applied loading, this result is due to the stress intensity factor definition where K depends on the applied stress and the size of the crack a, as per the relationship $K = \alpha\sigma\sqrt{\pi a}$. The numerical values are comparable to the analytical solution except that the XFEM method is more accurate compared to the FEM (Fig. 6.15), this fact is due to the enrichment of the finite elements field at the crack successively with the enrichment functions H and β_j. XFEM is therefore

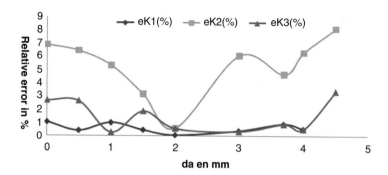

Fig. 6.15 Relative error of the XFEM and FEM compared to the theory

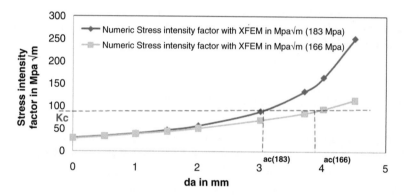

Fig. 6.16 Evolution of stress intensity factor versus applied loading and the advanced crack

more accurate than FEM thanks to the asymptotic enrichment term β_j at the crack. β_j improves the local description of the stress field σ so the FEM solution without enrichment β_j makes it difficult to approach the local characteristics of the crack tip solution.

We also note that XFEM with quadrangle elements is more accurate than XFEM with triangle elements, which is due to the fact that quadrangle elements in 2D and XC8R elements in 3D contain Gauss points higher than triangular elements TRI3 in 2D and TRI6 in 3D. The more Gauss points we have, the more the integration will be exact and the result will be more precise.

Figure 6.16 shows that the value a_c depends on the structure and the applied loading, for $\sigma = 166$ MPa: $a_c = 4$ mm equivalent to a final crack length $a_f = 17$ mm ($\frac{a_f}{W} = 0.85$) knowing that the initial crack $a_0 = 11$ mm ($\frac{a_0}{W} = 0.55$). Where a_c is the critical length of the crack advance corresponds to the critical stress intensity factor K_c leading to the fracture.

6.4.2 Comparison of CPU Times of Numerical Methods

The comparison of the process time for each numerical method is performed using the CPU time calculated by CAST3M Code. CPU time is the amount of time for which a central processing unit (CPU) was used for processing instructions in a computer program or operating system (Table 6.1).

Table 6.1 shows that the CPU calculation time in the XFEM using quadrangle elements (14,453 ms) is less compared to other numerical methods, namely FEM with quadratic elements (14,578 ms) and XFEM with triangular elements (16,147 ms), which proves that our modification of the integration strategy in XFEM is useful and has accelerated the execution of the system. The computation time of the FEM with quadratic elements is also comparable to that of XFEM with quadrangle elements, so the FEM is also a fast method, but it gives less precise results compared to XFEM, finally the integration with triangular elements in XFEM is slow but gives accurate results (Fig. 6.15).

6.4.3 Stress Concentration Coefficient K_t

In order to quantify the importance of the local stress increase, we calculated the values of the numerical stress concentration coefficient K_{tnum} along the axis of the 3D CT sample by the Peterson relationship [23] according to Eq. (6.41), then we compared it to the analytical value [24] according to Eq. (6.40):

$$K_{tanal} = 1 + 2\sqrt{\frac{a}{r}} \qquad (6.40)$$

$$K_{tnum} = \frac{\sigma_{max}}{\sigma_{nom}}, \qquad (6.41)$$

where

σ_{max}: the actual stress at the notch tip, σ_{nom}: the nominal stress observed in the far-field of a notch, a: depth of notch, r: radius of notch.

We have $a = 0.7$ mm and $r = 0.3$ mm. So the analytical value of K_{tanal} is 4.05.

Figure 6.17 shows a decrease in the value of the stress concentration coefficient K_t according to the horizontal axis of the CT sample, then a stabilization starting from 4 mm. The value obtained for K_t at the notch tip is maximum and is equal to 4.5 mm. This value is in accordance with that found analytically. The stress concentration coefficient far from the notch tends towards a limit value equal to 1.

Table 6.1 CPU times

CPU times (ms)					
Description	Operators number	Operators	XFEM_{QUADR}	XFEM_{TRI}	FEM_{QUA}
Mesh operators	26	OPTI	0	0	0
		FIN	0	0	0
		DENS	0	0	0
		DROI	0	0	0
		CERC	0	0	0
		INTE	0	0	0
		ET	0	0	0
		POIN	0	0	0
Boundary conditions…	7	PSCA	0	0	0
		REGL	0	0	0
		BLOQ	0	0	0
		TEXT	0	0	0
Math operators	43	MAXI	0	0	0
		>	0	0	0
		<	0	0	0
		OU	0	0	0
		EGA	0	0	0
		SIN	0	0	0
		COS	0	0	0
Processing of results	22	NON	15	15	0
		XTX	15	15	0
		MODE	15	15	31
		GRAD	15	15	187
		VOLU	31	62	46
		PROG	31	31	31
		SI	31	31	93
		INTG	46	62	93
		+	62	46	62
		ENVE	62	62	218
		CHAN	109	0	109
		WORK	109	93	46
		SIGM	140	140	140
		*	156	124	124
		ELIM	171	280	249
		MENA	187	187	140
		….	218	358	499
		PSIP	218	343	0
		ELEM	296	468	46
		RIGI	1388	1648	1123
		TRAC	2496	3697	2683
		RESO	8611	8424	8658
Sum	98		14,453	16,147	14,578

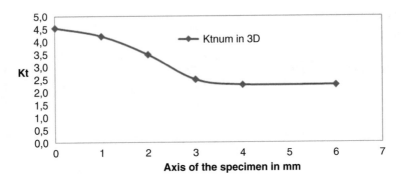

Fig. 6.17 Evolution of K_t along the axis of the CT sample

6.5 Conclusion and Perspective

This work presents the numerical modeling of the elastic behavior in 3D of a Compact Tension (CT) specimen in P265GH steel with extended finite element method (XFEM) by CASTEM code. We have simplified the definition of level sets and modified the integration strategy to facilitate the use of XFEM in CAST3M. The choice was to use quadrangle elements instead of triangle elements and to calculate the level sets from the crack mesh without the need for an intermediate grid to reproduce the advance of the crack. Harmfulness defect was modeled using the stress intensity factor. The evaluation of the stress intensity factors with XFEM and FEM and their comparison with the analytical solution was performed. The accuracy of XFEM using quadrangle elements compared to FEM and XFEM using triangle elements was highlighted. We also evaluated the stress concentration coefficient along the axis of the sample as well. On the basis of these simulations, we determined the value of the critical length leading to the fracture. We concluded that XFEM is an effective tool for modeling the crack growth in ductile materials which allows us to extend this work to simulate the elastic and elastoplastic problems of the pressure equipment when the crack path is not known a priori.

References

1. Fournier, D.: Analysis and Development of Refinement Methods hp in Space for the Neutrons Transport Equation, Doctoral Thesis University of Provence Marseille, (2011).
2. Shiozawa, K., Matsushita, H: Crack Initiation and Small Fatigue Crack Growth Behaviour of Beta Ti-15V-3Cr-3Al-3Sn Alloy, Proceeding Fatigue 96, G. Lütjering, H. Nowack (Eds.), Berlin, 301, (1996).
3. Tokaji, K., Takafiji, S., Ohya, K., Kato, Y, Mori, K.,: Fatigue Behavior of Beta Ti-22V-4Al Alloy Subjected to Surface-Microstructural Modification, Journal of Materials, (2003).
4. Broek D. Elementary engineering fracture mechanics. Dordrecht: Kluwer, (1991).
5. Gdoutos EE. Fracture mechanics - an introduction. Dordrecht: Kluwer, (1991).

6. El Hakimi, A. Le Grognec, P., Hariri, S.: Numerical and analytical study of severity of cracks in cylindrical and spherical shells, Engineering Fracture Mechanics, 1027–1044, (2008).
7. J.M. Melenk, I. Babuska, Comput. Methods Appl. Mech. Eng. 139, 289–314, (1996).
8. Pourmodheji, R., Mashayekhi, M.: Improvement of the extended finite element method for ductile crack growth. Department of Mechanical Engineering, Isfahan University of Technology, Isfahan 84156-83111, Iran Materials Science and Engineering journal (2012) homepage http://www.elsevier.com/locate/msea.2012
9. Belytschko, T., Black, T.: Int. J. Numer. Methods Eng. 45 601–620, (1999).
10. Moes, N., Dolbow, J., Belytschko, T.: Int. J. Numer. Methods Eng. 46, 131–150, (1999).
11. Dolbow, J.E.: Theoretical and applied mechanics, Ph.D. Thesis, Northwestern University, Evanston, IL, USA, (1999).
12. Moes, N., Sukumar, B., Moran, N., Belytschko, T.: Int. J. Numer. Methods Eng. 48, 1549–1570, (2000).
13. Samaniego, E., Belytschko, T.: Continuum-discontinuum modelling of shear bands. International Journal for Numerical Methods in Engineering, Vol. 62 (13), 1857–1872, (2005).
14. Elguedj, T.: Simulation numérique de la propagation de fissure en fatigue par la méthode des éléments finis étendus: Prise en compte de la plasticité et du contact-frottement. INSA de Lyon, (2006).
15. Benoit, P., Tamara, Y., Thierry, C., Simatos, A.: Propagation de fissures tridimensionnelles dans des materiaux inelastiques avec XFEM dans CAST3M. 10e colloque national encalcul des structures, May 2011, Giens, France. pp.Cle USB,<hal-00592709>, (2011).
16. http://www-CAST3M.cea.fr/.
17. Panetier, J.: Verification of stress intensity factors calculated by XFEM, PHD Thesis, The Normal School Of Cachan Superior, 36–40, (2010).
18. Barsoum: Further application of quadratic isoparametric elements to linear fracture mechanics of plate bending and general shells. Int. J. Num. Meth, Engng, 11, 167–169, (1976).
19. Singh, I.V., Mishra, B.K., Bhattacharya, S., Patil, R.U.: The numerical simulation of fatigue crack growth using extended finite element method, International Journal of Fatigue 36, 109–119, (2012).
20. Kumar, S., Singh, I.V. and Mishra, B.K., A coupled finite element and element-free Galerkin approach for the simulation of stable crack growth in ductile materials, Theoretical and Applied Fracture Mechanics, 70, 49–58, (2014).
21. Lahlou, M.: Numerical modeling and analytical validation of stress and stress intensity factor for SENT tensile sample of P265GH steel material, IPASJ International Journal of Mechanical Engineering (IIJME) Volume 3, Issue 4, ISSN 2321–6441, p 43, (2015).
22. Francois, D., Joly, L., La Ruprure Des Métaux, Masson et Cte. p 65, (1972).
23. Peterson, R.E.: Stress concentration factors, USA, John Willey et Sons, p 317, (1974).
24. Francois, Jol C.E.: Stress in a plate due to the presence of cracks and sharp corners, Trans Instn Nav. Archit, Vol. 55, p 219, (1913).

Part III
Computer Sciences and Smart Technologies

The third and final part of this volume is comprised of four chapters that can be viewed as applied mathematics to the broad area of "Computer Sciences and Smart Technologies". Chapter 7 written by Ilham Amezzane, Youssef Fakhri, Mohamed El Aroussi, and Mohamed Bakhouya gives an overview of a hardware acceleration of Supported Vector Machine training for real-time Embedded systems. In Chap. 8, Fatima-Zohra Hibbi, Otman Abdoun, and El Khatir Haimoudi present the requirements to integrate Artificial Intelligent (AI) in E-Learning Systems. Connected vehicles internet architecture is investigated by Zakaria Sabir and Aouatif Amine in Chap. 9. Finally, in Chap. 10, Amine Rghioui and Abdelmajid Oumnad describe a method using big data analysis for Internet of Things (IoT).

Chapter 7
Hardware Acceleration of SVM Training for Real-Time Embedded Systems: Overview

Ilham Amezzane, Youssef Fakhri, Mohamed El Aroussi, and Mohamed Bakhouya

Abstract Support vector machines (SVMs) have proven to yield high accuracy and have been used widespread in recent years. However, the standard versions of the SVM algorithm are very time-consuming and computationally intensive, which places a challenge on engineers to explore other hardware architectures than CPU, capable of performing real-time training and classifications while maintaining low power consumption in embedded systems. This paper proposes an overview of works based on the two most popular parallel processing devices: GPU and FPGA, with a focus on multiclass training process. Since different techniques have been evaluated using different experimentation platforms and methodologies, we only focus on the improvements realized in each study.

Keywords SVM · GPU · FPGA

7.1 Introduction

Many recent works highlight the importance of human activity recognition (HAR) applications in different areas of smart cities such as in healthcare and smart homes. For example, data collected from mobile devices to support healthcare diagnosis is a hot research topic, because it may facilitate health monitoring for elderly people at home, in real-time. While modern mobile devices are generally

I. Amezzane (✉) · Y. Fakhri · M. El Aroussi
LaRIT Lab, Faculty of Sciences, Ibn Tofail University, Kenitra, Morocco
e-mail: ilham.amezzane@uit.ac.ma; fakhri@uit.ac.ma; mohamed.elaroussi@ieee.org

M. Bakhouya
LERMA Lab, Faculty of Computing and Logistics, International University of Rabat, Sala Aljadida, Morocco
e-mail: Mohamed.bakhouya@uir.ac.ma

© Springer Nature Switzerland AG 2020
S. Dos Santos et al. (eds.), *Recent Advances in Mathematics and Technology*,
Applied and Numerical Harmonic Analysis,
https://doi.org/10.1007/978-3-030-35202-8_7

equipped with multi-core CPUs and graphics processing units (GPUs) that have powerful computation capabilities, complex tasks such as online machine learning (ML) are still challenging. Since they can help develop fast user-dependent and adaptive applications, HAR systems in some cases need to perform both the training and classification processes on the device itself. However, the training process is generally performed offline (i.e., on the server or the cloud) because traditional learning is based on large datasets, which is very time-consuming and power demanding. Therefore, needs of accelerating the training speed arise in order to meet time and resource constraints.

In a previous work [2], we conducted a comparison study exploiting feature selection (FS) approaches in order to reduce the computation and training time needed for the discrimination of six targeted physical activities while maintaining significant accuracy. The support vector machines (SVM) classifier offered the best compromise between accuracy and training time using different feature subsets. Moreover, with parallel execution over two CPU cores, we managed to reduce the number of data samples needed to reach the same significant training accuracy in less time than sequential execution.

Actually, SVMs have proven to yield high accuracy and have been used widespread in recent years especially for HAR applications. However, the training phase of SVM is a computational expensive problem, more particularly in online training problems where the time constraints are tight. Actually, sparse linear algebra, causing a huge computational load, is the main field of SVM, because the SVM solves the support vector by means of quadratic programming (QP) problem whose size is equivalent to the number of training samples. Due to the QP complexity, several decomposition methods have been proposed, among them is sequential minimal optimization (SMO). Nonetheless, conventional methods suitable for CPUs such as the well-known LibSVM, cannot be used for training large datasets in real-time embedded systems, as the available memory cannot store all elements of a kernel-value matrix and because intensive computations are both time and power consuming.

There are currently many accelerated libraries designed for multi-core CPUs and other accelerators. However, one of the increasingly popular trends in accelerating the SVM algorithm is the use of GPUs because they allow for a distribution of small single tasks among a large number of GPU cores, which should result in higher performance compared to CPU computation. Actually, GPU architecture is specialized for computer intensive, highly parallel computation, and therefore is designed such that more resources are devoted to data processing than caching and flow control. NVIDIA has developed their own programming framework for their GPUs, called Compute Unified Device Architecture (CUDA). It can be used to exploit the advantages of GPU architecture. On the other hand, OpenCL (Open Computing Language) framework is supported by AMD (CPUs, GPUs), Intel (CPUs, GPUs), Nvidia (GPUs that support CUDA), and Qualcomm (embedded/mobile CPUs), which make it a promising choice for parallelizing SVM because it allows the generated solution to be portable to a wide range of GPU manufacturers and allows making optimal use of the different computational components in one system.

Another alternative to CPUs is the use of field programmable gate array (FPGA) platforms. FPGAs are emerging in many areas of high performance computing, either as tailor-made signal processor, embedded algorithm implementation, software accelerator, or application specific architecture. Unlike traditional general processor, FPGA provides a facility of on-chip parallelism and pipelining. These two features highly increase the data throughput of FPGA-based system [1].

In this paper, we propose an overview of the works based on the most popular parallel processing devices: GPU and FPGA devices, with a focus on multiclass training process. Since different techniques have been evaluated using different experimentation platforms and methodologies, we only focus on the improvements realized in each study. The remainder of the paper is organized as follows: Sect. 7.2 reviews works on SVM using GPUs and FPGAs. Section 7.3 presents a comparison analysis on GPU and FPGA performances. Finally, conclusion and perspectives are presented in Sect. 7.4.

7.2 Research Works

7.2.1 Accelerating SVM Training with GPU

Catanzaro et al. [6] were the first who proved the effectiveness of implementing the modified SMO algorithm on GPUs. Their algorithm, called GPUSVM, was based on radial basis function (RBF) kernel. Evaluation was performed by examining a variety of different data structures. Authors reported a speedup in the range of 9–35 times for the training process and in the range of 81–138 times for the classification process.

In [9], Herrero et al. proposed the MultiSVM algorithm, where the potential of GPU computing was proved again. A multiclass SVM classifier based on the SMO algorithm was implemented. The training GPU model was realized by decomposition of the initial multiclass problem to many one-versus-all (OVA) classifications. The main achievement of the MultiSVM was the ability to execute all those processes in parallel over the same global memory. The algorithm was tested for the most popular datasets in SVM classification. Results showed dataset dependent speedups, while maintaining the accuracy, in the range of 3–57 times for training and 3–112 times for classification.

In [8], Cotter et al. presented a GPU-tailored SVM method for multiclass kernel SVMs, which can efficiently handle sparse datasets. Experimental results showed a speedup in the range of 38–78 times faster than CPU implementations, and in the range of 14–56 times faster than previous GPUSVM [6] implementations.

In [3], Athanasopoulo et al. proposed a modification of the LibSVM that pre-calculates the kernel matrix elements, using both CPU and GPU, in order to accelerate the training time without altering the performance. Experimental evaluation highlights how the GPU enables more efficient handling of large problems and shows higher impact than the CPU in the total processing time.

In [11], Li et al. introduced a new GPUSVM package which consists of a CUDA based parallel SMO. The core package includes a cross-validation (CV) tool, a fast training tool, and a predicting tool. Moreover, authors developed the algorithms in a way that they can be ported to other platforms such as OpenCL. GPUSVM showed better performance on medium to large datasets (binary and multiclass) as it achieved a speedup in the range of 2.27–77 times, compared to original LibSVM. However, the speed performance of CV was not done due to the difficulties of setting a base line.

Later, in [12], same authors proposed a novel parallel SVM training implementation to accelerate the CV procedure by running multiple training tasks simultaneously on a GPU. Therefore, redundant computations of kernel values across different training tasks are reduced. Consequently, the total time cost decreases significantly in the training process. Comparison tests have shown that the proposed method is 10–100 times faster compared to the state-of-the-art LibSVM tool.

In [16], Peters et al. proposed an accelerated implementation of SVM using a heterogeneous computing system programmed with OpenCL. The proposed framework was evaluated in terms of speed and accuracy for training and classification. The training was accelerated by a factor ranging from 9 to 22, and the classification by a factor of up to 12. This work only supports binary classification though.

In [5], Cagnin et al. proposed a GPU-based approach using OpenCL to improve the efficiency of parallelizing binary classification tasks. LibSVM's original source code computes the dot product as a double loop over the dataset objects, hence being of $O(N2)$ complexity, where N is the number of dataset objects. The proposed GPU-based parallel implementations compute dot products much faster than a CPU due to its stream processors. Results show that the proposed solution achieved a speedup of about 36 times the LibSVM's original version. This approach is capable of running in CPUs, GPUs, and even in mobile architectures.

In [7], Codreanu et al. described a novel approach for parallelizing multiclass SVM algorithm that converts a gradient-ascent CPU-based algorithm to an efficient GPU implementation. They have performed extensive comparisons between their algorithm and CPU and GPU implementations. The results of these comparisons show that their method is the fastest of all evaluated methods, especially if the examples consist of high-dimensional feature vectors.

In [13], Nan et al. proposed a modified LibSVM through the OpenCL framework for multiclass SVM. The approach speeds up the training time through improving the CV process. The proposed optimization method was applied in a mobile device without reduction in the accuracy rate. The parallel computing was 3.3 times faster than the serial computing in the PC, and 1.5 times faster in the mobile device.

In [22], Vanek et al. introduced a novel approach called Optimized Hierarchical Decomposition SVM (OHD-SVM) based on several known algorithms. The OHD-SVM implementation supports both dense and sparse datasets. Performance evaluation shows that dense datasets match better to the GPU architecture and that speedup over standard LibSVM may be significant. However, they performed only binary class RBF-SVM.

Finally, it should be noted that the most important limitations of the majority of the existing solutions in the literature are the use of: (1) dense matrix only format for storing datasets, (2) RBF only kernel without the possibility of changing the used kernel easily, and (3) binary only classification.

It is also worth noting that Google has recently introduced a framework called TensorFlow (TF) [21] which implements many ML algorithms using CUDA. Basically, TF was developed to perform computations based on the concept of data flow graphs. The graphs can be executed in parts or fully on available low-level devices such as CPU and GPU. TF offers other particularities: it is open source and can be implemented on Android smartphones. Moreover, TF implements a variant of SVM whose solution is obtained by a gradient descent, which is computed automatically (instead of a QP solver). Currently, only linear binary class SVMs are supported, but it is possible to code from scratch non-linear kernels and extend to multiclass.

7.2.2   Accelerating SVM Training with FPGA

FPGAs are reprogrammable hardware (HW) devices that offer an extremely promising source of HW acceleration. The power of FPGA lies in processing data in a parallel and pipelined way. Moreover, FPGA runtime reconfigurability allows the design to be scalable and adaptive to different types of input data [1]. In existing research works there are two typical approaches to speed up the SVM computations using FPGAs: (1) increase the level of parallelism by exploiting the inherent parallelism of the SVM algorithm, (2) reduce the bit width of the data representation which reduces the resource usage. In the following, we review some key works dealing with HW acceleration using at least one of the abovementioned approaches. We will focus on the multiclass SVM classifiers.

In [14], Papadonikolakis et al. proposed a scalable FPGA architecture which targets a geometric approach based on Gilbert's algorithm. The architecture is partitioned into floating-point and fixed-point domains in order to efficiently exploit the FPGA's available resources for the acceleration of the non-linear SVM training. Implementation results showed a speedup factor up to three orders of magnitude for the most computational expensive part of the algorithm, compared to a software implementation.

Later in [15], same authors focused on a performance comparison between a GPU implementation and the FPGA implementation of the previous work. Their motivation was to identify the most favorable one under different input configurations and resource constraints. The final speedup depended on the training set size.

Kuan et al. [10] proposed an architecture which consists of three main circuit modules functioning for the SMO process. The modules are a memory block and a cache block controlled by a designed finite-state machine (FSM) based controller. Experimental results showed a decrease in processing time from using the cache.

In [24], Wang et al. improved the modular design, proposed in [10], for realizing the computational bottleneck SMO on HW, while other processes were implemented in software. Compared to an ARM embedded C code, the proposed system achieved 90% reduction in training time with a slight decrease in classification performance.

Peng et al. [17] proposed a novel reconfigurable chip design for accelerating SMO-based SVM learning. Two novel methods were used in order to remove kernel cache design: (1) the first one targets the baseline SMO design, proposed in [10], by developing reconfigurable architectures with parallel and pipeline computing capabilities; and (2) the second one provides dynamic scheduling for an efficient reconfiguration. Simulation results achieved improvements in power consumption (17-fold improvement) and training speed (16-fold improvement) with satisfactory recognition accuracy (85%).

Shao et al. [19] introduced a novel optimized dataflow architecture that exploits the performance of incremental SVM training on FPGA. The proposed design is suitable for scenarios in which online SVM training is needed. Three challenges were addressed: random memory access, numerical accuracy, and list manipulation. Experimental evaluation achieved up to 40.97 times speedup against LibSVM software.

Zhao et al. [25] presented a multiclass SVM-based classification architecture that takes both FPGA acceleration approaches into account (i.e., reducing the bit width and increasing the parallelism). Moreover, the data type of the SVM model coefficients is configurable in order to support the trade-off between model accuracy and design parallelism. Performance evaluation results showed at least 14.2 times speedup and lower power consumption compared to two CPU platforms.

In [18], Bin rabieh et al. introduced FPGASVM framework, which aim is to tackle large-scale SVM problems that require frequent retraining. Ensemble learning is used to transform the overall training dataset into smaller datasets allowing each training subproblem to fully realize the parallelization potential. Moreover, cascaded multi-precision training flow is proposed by exploiting FPGA reconfigurability. Performance evaluation results showed that the system is an order of magnitude better than state-of-the-art CPU and GPU-based implementations, with low power consumption. This work supports only binary classification however.

Finally, Sirkunan et al. [20] proposed parameterizable linear kernel architecture in order to study the effect of varying the number of features and support vectors on the HW performance. Performance evaluation results showed that the number of features affects the maximum operating frequency, while the number of support vector affects the memory usage and the overall throughput.

7.3   GPU vs FPGA Performance Comparison

Although GPUs are cost efficient and benefit from shorter development time compared to FPGAs, the latter are power efficient. But taking a design decision is not straightforward. FPGAs are designed to perform fixed-point operations with

Table 7.1 Performance comparison: FPGA vs GPU [4]

Feature	Winner
Floating-point processing	GPU
Timing latency	FPGA
Processing/Watt	FPGA
Backward compatibility	GPU
Flexibility	GPU
Development time	GPU
Size	FPGA

a close-to-hardware programming approach taking massive benefit of bit-wise operations to maximize efficiency. GPUs on the other hand are designed for parallel processing of floating-point operations simplifying code adaptation from high-level programming languages. In addition, GPUs performance is measured in GFLOPS; while FPGAs processing power is measured in GMACS [4].

One of the most important limitations of existing embedded GPUs is the limited available resources (less power, memory, registers, cache, and cores). Accordingly, GPUs are difficult to be deployed in embedded environments, and this has motivated a move towards FPGA implementations. Therefore, power management techniques are extremely important to ensure longevity and reliability of GPUs in embedded systems. On the other hand, one of the most important limitations of existing embedded FPGAs is their significant HW development effort and consequently time-to-market. However, modern development tools that have been recently released allow simplified embedded systems design by using high-level languages. It can shorten time-to-market with no need for expert HW designers (e.g., Xilinx Vivado HLS tool [23]). Accordingly, FPGA offers a good promise for substantially accelerating SVM intensive computations with more flexibility at lower cost, while meeting hard constraints of embedded real-time systems. Table 7.1 summarizes this qualitative analysis for a faster understanding of the technology trade-offs.

7.4 Conclusion and Perspectives

In conclusion, both GPUs and FPGAs can offer significant improvements to the SVM training time. However, more research is required for FPGAs, taking into account the challenging trade-off between high classification accuracy and meeting real-time constraints. More research is also required for GPUs, taking into account the energy efficiency, which will have a crucial role in deciding their adoption in the context of embedded HAR systems. As far as we know, there is no work in the literature that implements GPU-based SVM training process for smartphone-based HAR. Therefore, in our ongoing research, we aim to experiment the TF framework, in order to facilitate the implementation of GPU-based multiclass SVM for online smartphone-based HAR, and in order to evaluate the training time, memory usage, and energy consumption in comparison to our previous multi-core CPU implementation.

References

1. Amezzane, I., Fakhri, Y., El Aroussi, M., Bakhouya, M.: FPGA Based Data Processing for Real-time WSN Applications: a synthesis. In: Proceedings of The First International Conference of High Innovation in Computer Science, pp. 83–86. ICHICS'16, Kenitra, Morocco (2016) http://www.uit.ac.ma/ichics2016/images/ICHICS%20Proceeding.pdf
2. Amezzane, I., Fakhri, Y., El Aroussi, M., Bakhouya, M.: Towards an Efficient Implementation of Human Activity Recognition for Mobile Devices. EAI Endorsed Transactions on Context-aware Systems and Applications. **18** (13): e3 (2018)
3. Athanasopoulos, A., Dimou, A., Mezaris, V., Kompatsiaris, I.: GPU acceleration for support vector machines. In Procs. 12th Inter. Workshop on Image Analysis for Multimedia Interactive Services (WIAMIS 2011), Delft, Netherlands (2011)
4. BERTEN: GPU vs FPGA Performance Comparison. WHITE PAPER, 19/05/2016 http://www.bertendsp.com/pdf/whitepaper. Cited 22 Nov 2018
5. Cagnin, H. E., Winck, A. T., Barros, R. C.: A Portable OpenCL-Based Approach for SVMs in GPU. Brazilian Conference on Intelligent Systems (BRACIS), pp. 198–203. Natal, Brazil (2015). https://doi.org/10.1109/BRACIS.2015.27
6. Catanzaro, B., Sundaram, N., Keutzer, K.: Fast support vector machine training and classification on graphics processors. In: Proceedings of the 25th international conference on Machine learning, pp. 104–111. ICML'08, ACM, New York, NY, USA (2008)
7. Codreanu, V., Droge, B., Williams, D., Yasar, B., Yang, P., Liu, B., Dong, F., Surinta, O., Schomaker, L.R., Roerdink, J.B. Wiering, M.A.: Evaluating automatically parallelized versions of the support vector machine. Concurrency and Computation: Practice and Experience. **28**(7), 2274–2294 (2016)
8. Cotter, A., Srebro, N., Keshet, J.: A GPU-tailored approach for training kernelized SVMs. In: Proceedings of the 17th ACM SIGKDD conference, pp. 805–813. KDD'11 (2011) http://doi.acm.org/10.1145/2020408.2020548
9. Herrero-Lopez, S., Williams, J.R., Sanchez, A.: Parallel multiclass classification using SVMs on GPUs. In: Proceedings of the 3rd Workshop on General-Purpose Computation on Graphics Processing Units, pp. 2–11. GPGPU'10, ACM, New York, NY, USA (2010)
10. Kuan, T. W., Wang, J. F., Wang, J. C., Lin, P. C., Gu, G. H.: VLSI design of an SVM learning core on sequential minimal optimization algorithm. IEEE Transactions on Very Large Scale Integration (VLSI) Systems. **20**(4), 673–683 (2012)
11. Li, Q., Salman, R., Test, E. et al. centr.eur.j.comp.sci. 1: 387 (2011) https://doi.org/10.2478/s13537-011-0028-7
12. Li, Q., Salman, R., Test, E., Strack, R., Kecman, V.: Parallel multitask cross validation for support vector machine using GPU. Journal of Parallel and Distributed Computing. **73** (3), 293–302 (2013)
13. Nan, Y.Y., Li, Q.Z., Piao, J.C, Kim, S.D.: GPU-Accelerated SVM Training Algorithm Based on PC and Mobile Device. International Journal of Knowledge Engineering. **2** (4), 182–186 (2016)
14. Papadonikolakis, M., Bouganis, C.S.: A scalable FPGA architecture for non-linear SVM training. In ICECE Technology. FPT 2008. International Conference on, pp. 337–340. IEEE (2008)
15. Papadonikolakis, M., Bouganis, C.S., Constantinides, G.: Performance comparison of GPU and FPGA architectures for the SVM training problem. In Field-Programmable Technology. FPT 2009. International Conference on, pp. 388–391. IEEE (2009).

16. Peters, E.: High Performance Implementation of Support Vector Machines Using OpenCL (Doctoral dissertation, Rochester Institute of Technology) (2015)
17. Peng, C.H., Chen, B.W., Kuan, T.W., Lin, P.C., Wang, J.F., Shih, N.S.: REC-STA: Reconfigurable and efficient chip design with SMO-based training accelerator. IEEE Transactions on Very Large Scale Integration (VLSI) Systems, **22** (8), pp. 1791–1802 (2014)
18. Rabieah, M.B., Bouganis, C.S.: FPGASVM: A Framework for Accelerating Kernelized Support Vector Machine. In Workshop on Big Data, Streams and Heterogeneous Source Mining: Algorithms, Systems, Programming Models and Applications, pp. 68–84 (2016)
19. Shao, S., Mencer, O., Luk, W.: Dataflow design for optimal incremental SVM training. In Field-Programmable Technology (FPT), 2016 International Conference on, pp. 197–200. IEEE (2016).
20. Sirkunan, J., Shaikh-Husin, N., Andromeda, T., Marsono, M.N.: Re-configurable logic embedded architecture of support vector machine linear kernel. In Electrical Engineering, Computer Science and Informatics (EECSI), 2017 4th International Conference on, pp. 1–5. IEEE (2017)
21. Tomeo.P.: Introduction to Machine Learning with TensorFlow Homepage https://www.slideshare.net/PTomeo1/introduction-to-machine-learning-with-tensorflow. Cited 22 Nov 2018
22. Vanek, J., Michalek, J., Psutka, J.: A Comparison of Support Vector Machines Training GPU-Accelerated Open Source Implementations. arXiv preprint arXiv:1707.06470 (2017)
23. Vivado High-Level Synthesis http://www.xilinx.com/products/design-tools/vivado.html. Cited 22 Nov 2018
24. Wang, J., Peng, J., Wang, J., Lin, P., Kuan, T.: Hardware/software co-design for fast-trainable speaker identification system based on SMO. 2011 IEEE International Conference on Systems, Man, and Cybernetics, pp. 1621–1625 (2011)
25. Zhao, R., Luk, W., Niu, X., Shi, H., Wang, H.: Hardware Acceleration for Machine Learning. 2017 IEEE Computer Society Annual Symposium on VLSI (ISVLSI), 645–650 (2017)

Chapter 8
Integrating an Intelligent Tutoring System into an Adaptive E-Learning Process

Fatima-Zohra Hibbi, Otman Abdoun, and El Khatir Haimoudi

Abstract With the emergence of new technologies, a new way of learning called "e-learning" has appeared, based on interactive training. This helps learners to develop their skills, while making the learning process independent of time and place. Artificial intelligence (AI) techniques in education were tested to increase learning experience. AI researchers look for domain applications, beginning to apply their ideas by tracking students on a course. Intelligent systems apply artificial intelligence techniques to meet the needs of their users; these methods allow the performance of an e-learning system to be increased by analyzing and adapting the profile of learning behavior and learning styles. Therefore, the objective of this chapter is to present the requirements for integrating AI into e-learning systems. The results showed significant advantages when including AI techniques, specifically intelligent tutoring system, in the learning process.

Keywords Artificial intelligence · E-learning system · Learning process · Intelligent tutoring system

8.1 Introduction

In classical e-learning, we observe that the learner is isolated and depressed with regard to the system, because there is no interaction in real time between the learner and the applications. The learner is passive according to the simple page, which contains a passive course and ends with a list of questions or exercises. For that reason, most learners have a negative feeling during or after their use of the system. Another point is that the system cannot extract the needs of learners; along with the problem of inelasticity, there is an absence of two very important

F.-Z. Hibbi (✉) · O. Abdoun · E. K. Haimoudi
Laboratory of Advanced Sciences and Technologies, Polydisciplinary Faculty, Abdelmalek Essaadi University, Larache, Morocco

© Springer Nature Switzerland AG 2020
S. Dos Santos et al. (eds.), *Recent Advances in Mathematics and Technology*,
Applied and Numerical Harmonic Analysis,
https://doi.org/10.1007/978-3-030-35202-8_8

things in the learning process: the flexibility and learning between teacher and learner. The e-learning system comprises intelligent tools for analysis, evaluation, and evaluation of the user's knowledge and skills in addition to the monitoring and supervision of the e-learning process. AI allows use of its techniques to implement better performing education systems, such as the genetic algorithm, the intelligent tutoring system, and neural networks. Environments that use teachable agents and animated interfaces encourage and motivate student learning. The main challenges currently facing the use of educational agents and systems concentrate on how to make them useful and how best to include them in the learning experience [1]. In this regard, we propose the integration of an intelligent tutoring system into an adaptive e-learning process. This integration provides immediate and customized feedback, and it can also measure students' motivations produced to attract their interest. Furthermore, it can resolve the problem of inelasticity and adaptability of the learner's requirements, replacing them with flexibility and analysis, which can investigate the student's strategies.

8.2 Limitations of a Classical E-Learning

E-learning is not a panacea for training, a positive feature of e-learning is that it is virtual, but this can also be negative: it prevents any human contact face to face. For those who work in an open space, this does not pose a problem a priori, but for people who are already isolated, this may be difficult. Faced with the large abandonment of modules, some participants may be afraid, or not dare to use them, or not know how to use them, and therefore spend more time on this technical aspect. The real problem is to adapt the content depending on the type of audience. It is usually created in advance. However, any trainer who intervenes knows that he will have to adapt the vocabulary or deepen this or that part, depending on his audience. In e-learning, it is fixed. Figure 8.1 presents the limitations of classical e-learning.

8.3 Intelligent Tutoring System

8.3.1 Definition

An intelligent tutoring system (ITS) is a system that provides feedback on the actions of learners without the intervention of a human being. It provides learners with the opportunity to practice their skills by executing tasks in highly interactive learning environments. On the other hand, an ITS evaluates the actions of each learner in these interactive environments and develops a model of their knowledge, skills, and expertise. Using the learner model, it can adapt teaching strategies,

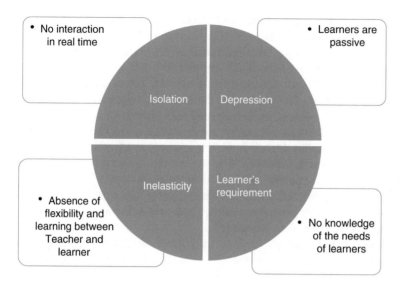

Fig. 8.1 Limits of classical e-learning

Fig. 8.2 Intelligent tutoring
system

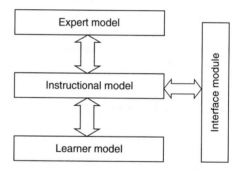

with regard to either content or style, and provide explanations, advice, examples, demonstrations, and concrete problems relevant to the individual learner.

8.3.2 Architecture

In this section, we present the architecture of an ITS and its components. Figure 8.2 shows the interface of a model ITS. It consists of an expert model, an instructional model, and a learner model. An ITS consists of four different modules.

The interface module is important for use as a method of communication and as a learning environment that can help the learner with a task. It can also provide an external representation of the expert model and the pedagogical model. These kinds of tutoring systems can provide the learner with a large selection of practice database case studies along with individualized feedback, to solve each case study. In addition, it is very practical for learners who need to practice and learn at their own speed [5] and problem-solving capacity (procedural knowledge). This knowledge allows the ITS to compare the learner's actions and selections with an expert's to evaluate what he or she does and does not know [3, 5].

The expert model is a computation representation of the knowledge of an expert in the field (declarative knowledge). The learner model is a state of knowledge of the learner as he/she interacts with the tutoring system. The model evaluates each learner's performance based on their behavior while interacting with the tutoring system to determine their knowledge, perceptual capacities, and reasoning capacities. The model generates evidence and uses inference to give a number of relevant instructions to each learner [5].

The instructional model (pedagogical) model includes knowledge to make decisions about teaching techniques. It is based on the diagnostic processes of the learner's model to make decisions. For example, if a learner has been considered a beginner in a specific procedure, this model presents step-by-step demonstrations of the procedure before asking the user to perform the procedure on his or her own. When a learner obtains expertise, this model may decide to stage increasingly complex scenarios. In addition, this model can choose topics, simulations, and examples appropriate to a learner's level of knowledge [3].

8.3.3 Application of ITS

In this part, we illustrate some applications of ITS such as an "adaptive intelligent tutoring system for an e-learning system." This work is based on the combination of an ITS and adaptive hypermedia and was a natural starting point for research on adaptive educational hypermedia into adaptive intelligent tutoring systems [5]. The second application is "Using a behavioral analysis system in an ITS." The aim of this work is to integrate a behavioral analysis system that provides the creation of a profile that includes learners with the same behavior in a student model of the ITS. The goal was to help the pedagogical assistance to create more specification for all learners even though others did not pass the test [4]. The third application is "integrating affect sensors into an intelligent tutoring system." The objective of this work is to develop an agile learning environment that is sensitive to a learner's affective state and to integrate state-of-the-art, non-intrusive, affect-sensing technology with Auto Tutor in an endeavor to classify emotions on the bases of facial expressions, gross body movements, and conversational cues.

8.4   Related Work

8.4.1   Implicit Strategies for Intelligent Tutoring Systems

The implicit strategies of a new approach based on an unconscious process that addresses the automatic mechanisms associated with learning and cognitive processing. This integration provides tutors with new implicit strategies relying on indirect interventions. This strategy integrates more complex coaching techniques using realistic characters, allowing real-time emotional interactions between virtual pedagogical agents and learners. Figure 8.3 describes the proposed approach, which includes the exciting ITS, and an implicit tutor is added, based on implicit interventions and the use of an implicit toolbox that contains a subliminal perception, an interface interaction. Figure 8.3 presents the model of an implicit strategy.

8.4.2   Explicit Strategies for Intelligent Tutoring Systems

There are different techniques of direct intervention. The first strategy is based on a variety of tasks and support, and uses a form of examples or definitions to help the student to understand specific concepts. The second strategy uses techniques addressing frustrated and bored learners to measure their stress levels by skin conductance. Burleson provides an agent that uses the learner's facial expressions and movements. For example, he can smile if he sees the learner smiling [9] (Fig. 8.4).

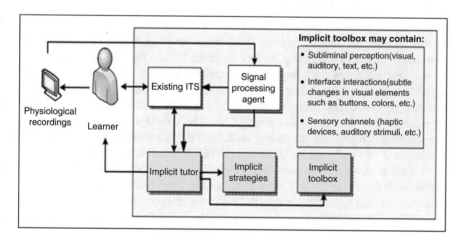

Fig. 8.3 Model of an implicit strategy for ITS [8]

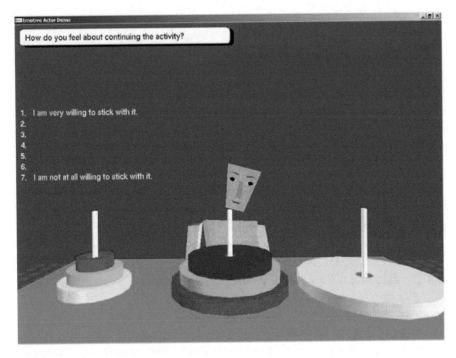

Fig. 8.4 Mimics agent

8.5 Model of Smart Tutoring

8.5.1 Proposed Approach

In this section, we present our model of smart tutoring, which is based on the integration of implicit and explicit interventions in an ITS. We are interested in a tutor that integrates both implicit and explicit strategies. This tutor selects the appropriate strategy according to three criteria: the learner's profile, data from learning progress, and affective reactions. Figure 8.5 illustrates our proposed approach, which includes the existing ITS. The development of this approach involved two components: the student model and the instructional model. This latter is where we will make the decision and use one of these tutors (implicit/explicit). Each tutor uses his or her own strategies and tool box. The selection of one of the tutors is based on the following criteria: data from learning progress (we can extract these data from the pedagogical tool); the learner's profile (we can use the integration of a behavioral analysis system into the ITS); and affective reactions (with interaction of a stimulus evaluation check system).

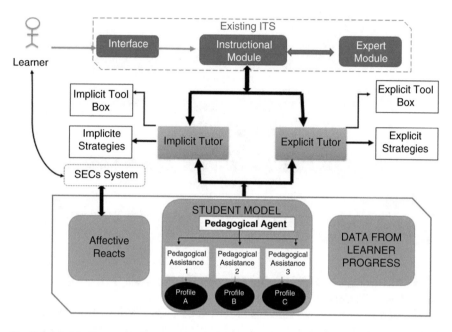

Fig. 8.5 Model of smart tutoring

8.6   Smart Tutoring System: Developed Platform

In this part, we present the context of the developed platform, the proposed solution, and the result, which can be the integration of our proposed approach into the smart tutoring system.

8.6.1   The Context and Proposed Solution

The aim of this smart tutoring system is to improve high-quality learning, to facilitate communication between the teacher and the learner, to track student progress, and to provide advice, guidance, and feedback. In this regard, we propose a smart tutoring system. This system is an e-learning space characterized by its portability and compatibility with different media (tablet, smartphone, and computer), and is aimed at learners and teachers. The space is endowed by several collaborative workspaces, including:

The teacher space allows teachers to create courses, drop off their course materials and practical and/or directed assignments in different formats (PowerPoint, PDF, World, etc.); thus, it contains a white examination creation area and space to enter

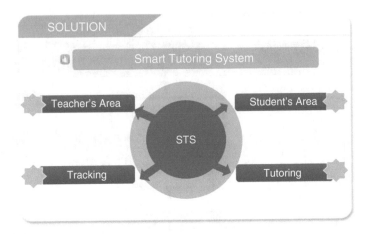

Fig. 8.6 Proposed solution

notes. This workspace also includes a section for discussion between the teachers and students or teachers among themselves.

The student space allows students to consult the course, write their (TD/TP), and to take a mock examination to verify his or her understanding and to ensure that he or she prepares for the final examination. It also contains a space for communication with their teachers (Fig. 8.6).

8.6.2   Results

The model of smart tutoring that we are developing will be applied in the e-learning platform of the Polydisciplinary Faculty of Larache. We chose to use the "Moodle" platform because it is one of the most popular open-source platforms in the world, with its large Francophone and international community. It is clear, well-structured, useful documentation, and its many discussion forums are focused on all the problems generated by the complexity of distance learning. Moodle is extremely modular, it has several collaboration tools and a good follow-up to training, statistics, multi-criteria reporting, and it supports several languages, even Arabic. In Fig. 8.7 we present the consulting interface of the course, which contains several pages and is presented as a slide. The student can also add notes or remarks on the content. In Fig. 8.8, we illustrate the passing interface of the examination. The student take this type of examination in real mode. This means that the teacher must determine the time and date of the examination so that all students are connected. If everyone is ready, the teacher launches question after question and each question is of limited duration. Access to this type of examination requires the presence of the teacher in connected mode.

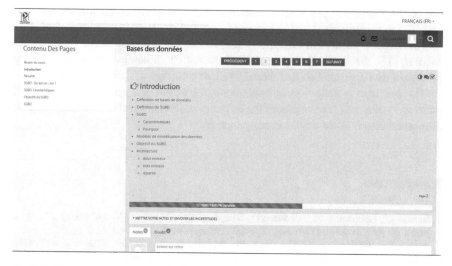

Fig. 8.7 The consulting interface of the course

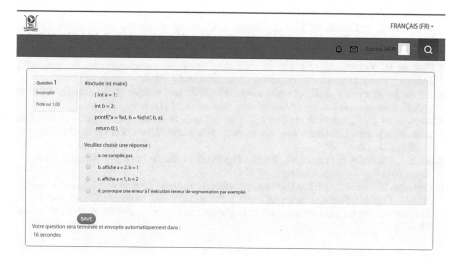

Fig. 8.8 The passing interface of the examination

8.7   Conclusion and Perspectives

Intelligent e-learning systems are knowledge-based systems that imitate the human mind. The principal characteristics of these systems are inference capacity, reasoning, perception, and learning [2, 3]. This system provides the student with an automatic learning process and a list of suitable activities. Thus, we can say that ITS are different from traditional methods because while we study, the learning is not only by listening and writing passively, but also provides significant learning activities in an individual and independent way [6, 7].

This paper presents the limitations of classical e-learning systems, analyzing an existing ITS, the synthesis work of implicit/explicit strategies for an ITS and we propose a new approach named model of smart tutoring.

The next step is the practical implementation of the new approach in our smart tutoring system and integration of the learning analytical mechanisms into the first desired criterion: "Data from learner progress".

References

1. Biswas, G., Leelawong, K., Schwartz, D., Vye, N. (2005). *"Learning by teaching: a new agent paradigm for educational software"*. *Applied Artificial Intelligence 19(3–4), 363–392*.
2. de los Angeles Constantino-González, M., Suthers, D. D. (2000). *"Coaching web-based collaborative learning based on problem solution differences and participation"*. *International Journal of Artificial Intelligence in Education 13, 324–333*.
3. Mihalca, R., Andreescu, A. (2008). *"Knowledge management in e-learning systems"*. *Revista Informatica Economică 2 (46), 60–65*.
4. El Haddioui, I., Khaldi, M. (2011). *"Behavioral analysis of learners on an online learning platform"*, *the 7th International Scientific Conference eLearning and Software for Education Bucharest*.
5. Phobun, P., Vicheanpanya, J. (2010). *"Adaptive intelligent tutoring system for e-learning systems"*. *Procedia Social and Behavioral Sciences 2, 4064–4069*.
6. Felder, R., Spurlin, J. (2005). *"Applications, reliability, and validity of the index of learning styles"*, *International Journal of Engineering Education. 21 (1), 103–112*.
7. Johnson, W. L., Rickel, J. W., Lester, J. C. (2000). *"Animated pedagogical agents: face-to-face interaction in interactive learning environments"*, *International Journal of Artificial Intelligence in Education, 11, 47–78*.
8. Jraidi, I., Chalfoun, P., Frasson, C. (2011). *"Implicit Strategies for Intelligent Tutoring Systems"*, *Springer-Verlag Berlin Heidelberg*
9. Burleson, W. (2006). *"Affective learning companions: strategies for empathetic agents with real-time multimodal affective sensing to foster meta-cognitive and meta-affective approaches to learning, motivation, and perseverance"*, *Media Arts and Sciences, School of Architecture and Planning*

Chapter 9
NDN vs TCP/IP: Which One Is the Best Suitable for Connected Vehicles?

Zakaria Sabir and Aouatif Amine

Abstract Current Internet architecture is based on the TCP/IP model and has been developed several years ago. So far, it is the main architecture used in different fields of science and technology. This architecture however may need to be reconstructed to fulfill future aims and domains, such as connected vehicles, which is our field of study. For this reason, we suggest in this paper a comparison of the current architecture with NDN (named data networking), a future Internet architecture, in order to help the reader distinguish the characteristics of the two architectures, and thus choosing the best one depending on his field of study.

Keywords NDN · TCP/IP · CCN · ITS · V2V · V2I · Connected vehicles · Road safety

9.1 Introduction

In the last few years, improving road safety using science and technology is considered as one of the most important issues. Thus, researchers have become interested in intelligent transportation systems (ITS) as a promising way to reduce the accident rate. In this context, the connected vehicle system, which is one of the most important components of ITS and smart technologies, has been proposed as a hopeful technology. This technology uses wireless communication and brings into focus supporting mobility, safety, and environmental applications used, for example, in smart cities.

Most connected vehicle applications are IP-based network protocol, which influence both quality of service (QoS) and latency. Nevertheless, the IP protocol is designed for host-to-host connection which is low effective for information dissem-

Z. Sabir (✉) · A. Amine
BOSS Team, LGS Laboratory, ENSA of Kenitra, Ibn Tofail University, Kenitra, Morocco
e-mail: zakaria.sabir@uit.ac.ma; aouatif.amine@uit.ac.ma

© Springer Nature Switzerland AG 2020 151
S. Dos Santos et al. (eds.), *Recent Advances in Mathematics and Technology*,
Applied and Numerical Harmonic Analysis,
https://doi.org/10.1007/978-3-030-35202-8_9

ination. Moreover, IP protocol is unsatisfactory in supporting direct communication in connected vehicles with high mobility and in absence of Road Side Units (RSU).

Recently, many researchers begin to test NDN in connected vehicles applications. NDN belongs to content-centric networking (CCN). Its concept suggests focussing on the content (what to send) rather than the address (where to send). This new approach supports the caching of data in the network in order to satisfy future requests. This gives an advantage to NDN in terms of data transfer.

This work presents a comparative study between TCP/IP (Transmission Control Protocol/Internet Protocol), the current Internet model, and NDN, the future Internet architecture. It attempts to show basic functioning components of both the systems in a simplified manner. By the end of this paper, the most important differences of the two architectures will be clear to the reader. To be noted this work is financed by the ministry of equipment transport logistics and water in collaboration with the national center of scientific and technical research under the project named "SafeRoad: Multi-platform for Road Safety (MRS)."

The remainder of the paper is organized as follows: Sect. 9.2 discusses the fundamental limitations of the current Internet architecture. Section 9.3 is devoted to a presentation of NDN. Our comparative study is presented in Sect. 9.4. Finally, Sect. 9.5 concludes the paper.

9.2   Limitations of the Current Internet (TCP/IP)

The TCP/IP is the current Internet model providing end-to-end connectivity and specifying the transmission, routing, addressing, formatting, and receiving of data at the destination. IP is handling datagram routing while TCP is responsible for higher-level functions like error detection and segmentation [1].

However, research has consistently shown that the current Internet architecture has limitations. Researchers of FIA (Future Internet Architecture) discussed fundamental limitations of this architecture [2, 3] in terms of processing/handling of data, storage, transmission, and control.

In processing/handling of data, which refers to forwarders, computers, CPUs, etc., hosts are unable to run appropriate actions and the failures are not identified. There is also an absence of some useful services like transportation and health care. In storage, which refers to disks, buffers, caches, memory, etc., problems like availability of information while transferring, storage management and retrieval, loss of integrity and storage encrypted data, are faced due to attacks and breakdowns. Transmission of data, which refers to the exchange of data, also suffers from problems of security, since it is not built in the architecture, but only given by different extensions. In control and supporting mobility (which we are interested in), that refers to analysis, observation, and decision, there is a lack of congestion control, as current schemes are based on collaboration between the network and end systems which causes more expenses. We have recapitulated the limitations of the TCP/IP architecture in Fig. 9.1.

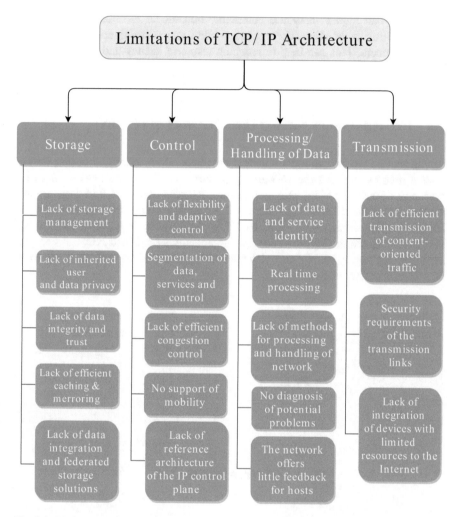

Fig. 9.1 Limitations of the current Internet architecture according to researchers of FIA

Shang et al. analyzed some challenges in IoT networking via TCP/IP architecture in different points [4]. They indicated that the design of the TCP/IP protocol stack is not a good fit for the IoT environment, since it faces a lot of issues, in terms of the network layer, transport layer, and application layer.

The TCP protocol cannot support a variety of communication patterns. Devices may activate sleep mode due to the energy constraints; therefore, the point-to-point connection cannot continue to be maintained. Another difficulty is caching. The model of TCP/IP necessitates that both the client and the server are connected at the same time, which is challenging due to the intermittent and dynamic network environment. The IoT applications (like connected vehicles), however, are based on caching and request–response communication model in order to accomplish effective diffusion of data. The cached content can serve future requests from other consumers. This can help in reducing response latency and saving network bandwidth.

9.3   Presentation of Named Data Networking (NDN)

In this section we will introduce named data networking briefly, focusing on the system architecture and the forwarding process. The NDN is considered as an evolution of the current Internet architecture. Its main idea is to find data through naming data packets instead of source and destination address [5–7]. Instead of the IP address, the name prefix is used by NDN-based routers to forward packets [8]. Figure 9.2 depicts the layered hourglass architectures of IP and NDN. As can be seen from the figure, NDN changes the global ingredient of the network stack from IP to blocks of named content [9].

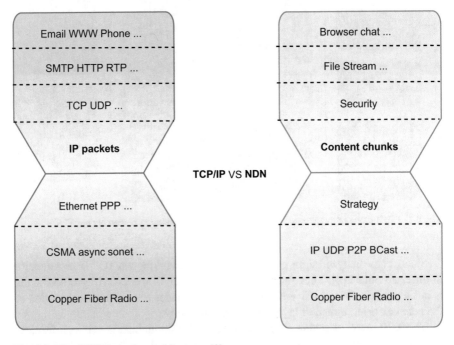

Fig. 9.2 IP vs NDN hourglass architectures [9]

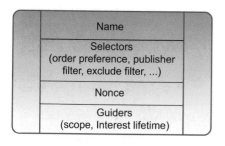

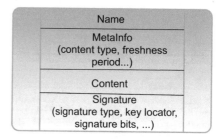

Interset packet   **Data packet**

Fig. 9.3 Interest and Data packets in the NDN architecture [14]

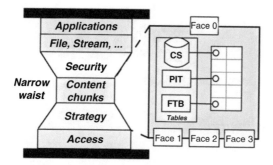

Fig. 9.4 Forwarding structures in NDN [15]

9.3.1   NDN System Architecture

Two different types of packets are used in NDN communication: "Interest packet" and "Data packet" [10] as illustrated in Fig. 9.3. In order to get a specific content, a consumer broadcasts an Interest packet, that carries the name of the desired piece of data, over the available network interfaces. This name is used by the routers to forward the packet in the network [11]. Once the Interest reaches the provider, i.e., the original data owner or any other node that keeps a cached copy, will replay with a Data packet that contains both the name and the content [12] in addition to a signature by the producer's key [13]. To get back to the requesting consumer, the Data packet will pursue in reverse the path taken by the Interest.

The processing of Interests is done using the following structures that are managed by every NDN node, as depicted in Fig. 9.4: the CS (content store) which is responsible for caching incoming Data, the PIT (Pending Interest Table) that stores the Interests forwarded by the router and still not satisfied yet, and the FIB (forwarding information base) which is populated by a specific routing protocol and used to forward Interest packets towards possible corresponding Data sources.

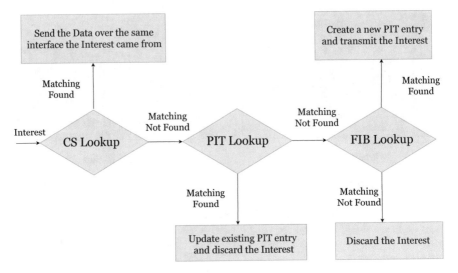

Fig. 9.5 Forwarding process at an NDN node [16]

9.3.2 Forwarding Process

When an Interest packet arrives at an NDN node, the node runs the following algorithm in order to find the desired Data, this is illustrated in Fig. 9.5:

- First, it checks the CS. If there is a matching entry, the node replays with a Data packet, using the same interface the Interest came from.
- Else, it looks in the PIT. If a matching entry is found, the existing PIT entry is updated with the Interest packet's incoming interface and the Interest is discarded.
- If the matching was not found in the PIT, the content is looked up in the FIB through performing a longest prefix match. If a matching FIB entry exists, the Interest is transmitted to the specified outgoing interface(s) and a new PIT entry will be created for the packet. Otherwise, depending on the forwarding policy, the interest will be forwarded to all outgoing interfaces or deleted.

When the data arrives, the NDN node forwards it to the interfaces recorded in the corresponding PIT entry, caches the data in its CS, and removes that PIT entry.

9.4 Comparative Study

In this section, we will present a comparative study of the two models: TCP/IP (current Internet architecture) and NDN (future Internet architecture) in a simplified way, in order to determine the best suitable architecture for connected vehicles.

The TCP/IP model has a host-to-host communication pattern. It maintains the FIB entity and identifies the end host using IP prefixes. NDN is based content-centric, uses name prefixes, and maintains three entities: CS, PIT, and FIB. In Internet, packets are not registered in the routing table, while in NDN, routers save a copy of each Interest packet transiting through them until the expiry of the packet's lifetime.

In the TCP/IP, using congestion management schemes explicitly is needed, since packets can pursue any direction to reach the destination, causing congestion in dynamic environments. But in NDN, adjusting packet rate by permitting one Data packet for each Interest packet could solve this problem. NDN routers use nonce with packet name to avoid loops, whereas TCP/IP routers rely on routing protocols. To provide security, a secure medium is used between hosts on the Internet, while NDN include security in the architecture, as every Data packet is signed by the original provider.

While TCP/IP cannot monitor data rendering, since the packets are not recorded in the routing table, NDN can do it thanks to PIT state and calculation of round trip time (RTT). In-network caching is not supported on the Internet, whereas, NDN architecture support caching of the solicited data. In NDN, examining interfaces regularly, allows controlling failures in packet forwarding. In the TCP/IP, this is done via routing protocols which use coherent routing tables. However, this may generate extra charges in the network. We summarized the characteristics of each architecture as can be found in Table 9.1.

Table 9.1 Comparison of TCP/IP and NDN

Characteristics	TCP/IP	NDN
Communication pattern	Host-to-host	Content-centric
Identification of end host	IP address	Content name
Type of packets	IP packet	Interest and Data Packets
Supporting caching	No	Yes
Use of DNS	Mandatory	Not needed
Packets registration	No	Yes
Entities	FIB	CS, PIT and FIB
Packet failures management	Yes	Yes
Forwarding packets	IP prefixes	Name prefixes
Type of connection	Point-to-point	Multipoint-to-multipoint
Monitoring data rendering	No	Yes
Information dissemination	Ineffective	Wide-ranging
Congestion management	Using some schemes explicitly	Adjusting Interest packet rate
FIB storage	Next hop information only	Multiple hope status
Avoiding loops	Routing protocol	Nonce and packet name
Security	Securing channel	Inherent

9.5   Conclusion

The aim of this work is to give a clear comparison of two Internet architectures TCP/IP and NDN. This comparison will help the reader making sense of the future Internet. From the outcome of our comparative study, it is possible to conclude that named data networking appears to be the best suitable approach for connected vehicles, since it has the advantage in different points, especially supporting in-network caching and providing an architecture with inherent security.

On the basis of the comparisons presented in this paper, work on the remaining issues is continuing and will be presented in future papers. We will conduct a study in order to identify the consequence of all the characteristics that we compared in the last part of this paper, in terms of cost and security for connected vehicles end-users.

Acknowledgements This research work is supported by the "SafeRoad: Multi-platform for Road Safety (MRS)" Project under Contract No: 24/2017, financed by the Ministry of Equipment Transport Logistics and Water in collaboration with the National Center of Scientific and Technical Research.

References

1. Forouzan, B.A.: TCP/IP Protocol Suite. McGraw-Hill Higher Education (2002).
2. Izquierdo, Stamoulis, Q.G., Alvarez, A.F.J., Melideo, U.M., Niccolini, S., Hauswirth, N.M., Krummenacher, D.R.: Fundamental Limitations of current Internet and the path to Future Internet 1 EC. Presented at the (2011).
3. NSF Future Internet Architecture Project, http://www.nets-fia.net/, last accessed 2018/03/21.
4. Shang, W., Yu, Y., Droms, R.: Challenges in IoT Networking via TCP/IP Architecture. 7 (2016).
5. Zhang, E., Burke, J., Thornton, S., Zhang, T., Claffy, K., Massey, P., Abdelzaher, W., Crowley, Y.: Named Data Networking (NDN) Project. 27 (2010).
6. Zhang, L., Estrin, D., Burke, J., Jacobson, V., Thornton, J., Uzun, E., Zhang, B., Tsudik, G., Claffy, K., Krioukov, D., Massey, D., Papadopoulos, C., Ohm, P., Abdelzaher, T., Shilton, K., Wang, L., Yeh, E., Crowley, P.: Named Data Networking (NDN) Project 2011–2012 Annual Report.
7. Jacobson, V., Burke, J., Estrin, D., Zhang, L., Zhang, B., Tsudik, G., Claffy, K., Krioukov, D., Massey, D., Papadopoulos, C., Ohm, P., Abdelzaher, T., Shilton, K., Wang, L., Yeh, E., Uzun, E., Edens, G., Crowley, P.: Named Data Networking (NDN) Project 2012–2013 Annual Report.
8. Yi, C., Afanasyev, A., Moiseenko, I., Wang, L., Zhang, B., Zhang, L.: A case for stateful forwarding plane. Comput. Commun. 36, 779–791 (2013).
9. Jacobson, V., Smetters, D.K., Thornton, J.D., Plass, M.F., Briggs, N.H., Braynard, R.L.: Networking Named Content. In: Proceedings of the 5th International Conference on Emerging Networking Experiments and Technologies. pp. 1–12. ACM, New York, NY, USA (2009).
10. Hassan, S., Habbal, A., Alubady, R., Salman, M.: A Taxonomy of Information-Centric Networking Architectures based on Data Routing and Name Resolution Approaches. J. Telecommun. Electron. Comput. Eng. JTEC. 8, 99-107-107 (2016).
11. Bari, M.F., Chowdhury, S.R., Ahmed, R., Boutaba, R., Mathieu, B.: A survey of naming and routing in information-centric networks. IEEE Commun. Mag. 50, 44–53 (2012).

12. Amadeo, M., Campolo, C., Molinaro, A.: A novel hybrid forwarding strategy for content delivery in wireless information-centric networks. Comput. Commun. 109, 104–116 (2017).
13. Wu, T.-Y., Lee, W.-T., Duan, C.-Y., Wu, Y.-W.: Data Lifetime Enhancement for Improving QoS in NDN. Procedia Comput. Sci. 32, 69–76 (2014).
14. Zhang, L., Afanasyev, A., Burke, J., Jacobson, V., Claffy, KC, Crowley, P., Papadopoulos, C., Wang, L., Zhang, B.: Named Data Networking. SIGCOMM Comput Commun Rev. 44, 66–73 (2014).
15. TalebiFard, P., Leung, V.C.M., Amadeo, M., Campolo, C., Molinaro, A.: Information-Centric Networking for VANETs. Veh. Ad Hoc Netw. 503–524 (2015).
16. Amadeo, M., Campolo, C., Molinaro, A.: Information-centric networking for connected vehicles: a survey and future perspectives. IEEE Commun. Mag. 54, 98–104 (2016).

Chapter 10
Predictive Analysis for Diabetes Using Big Data Classification

Amine Rghioui and Abdelmajid Oumnad

Abstract The Internet of Things (IoT) relies on physical objects interconnected between each other's, creating a mesh of devices producing information. In this context, sensors are surrounding our environment (e.g., Healthcare's, buildings, and smartphones) and continuously collect data about our living environment. The explosive growth in the number of devices connected to the Internet of Things (IoT) in smart Healthcare only reflects how the growth of big data perfectly overlaps with that of IoT. Predictive analytics helps the physicians, doctors to identify the patient admission to hospital at an early stage. To perform predictive analytics in the health field, several factors must be considered: demographic data, hospital parameters, patient history and several other indicators for specific diseases. In this paper, we propose a predictive model for diabetic patients using Naive Bayes, Random Forest, Naive Bayes, and j48 classification algorithm from the diabetes data set to test the most powerful to determine the patient's level of risk.

Keywords Internet of Things · Big data · Classification · Healthcare

10.1 Introduction

The Internet of Things (IoT) is a computing concept that describes a future where every day physical objects will be connected to the Internet and be able to identify themselves to other devices. Over the last two decades, we have seen an enormous amount of growth in data. Because of this technological revolution, the big data is becoming increasingly an important issue in the sciences. The term "Big Data" was first used in 1997 by Cox and Ellsworth in [1] to refer to "input data sets arising in scientific visualization that are quite large." A conservative definition given by IDC

A. Rghioui (✉) · A. Oumnad
Research Team in Smart Communications-ERSC-Research Centre E3S, EMI, Mohamed V University, Rabat, Morocco
e-mail: rghioui.amine@gmail.com; aoumnad@emi.ac.ma

© Springer Nature Switzerland AG 2020
S. Dos Santos et al. (eds.), *Recent Advances in Mathematics and Technology*,
Applied and Numerical Harmonic Analysis,
https://doi.org/10.1007/978-3-030-35202-8_10

is that BD refers to "a new generation of technologies and architectures designed to economically extract value from very large volumes of a wide variety of data by enabling high-velocity capture, discovery, and/or analysis." With the increasing cost for healthcare services and increased health insurance premiums, there is a need for proactive healthcare management and wellness. This shift from reactive to proactive healthcare can result in improved quality of care, decrease in healthcare costs, and eventually lead to economic growth. In recent times, technological breakthroughs have played a significant role in empowering proactive healthcare. For instance, real-time remote monitoring of vital signs through embedded sensors (attached to patients) allows health care providers to be alerted in case of an anomaly. Furthermore, healthcare digitization with integrated analytics is one of the next big waves in healthcare. The goal is to understand population health for disease control and predictive analysis. For instance, predictive analysis can help understand aggravating health conditions and could prevent adverse health events from occurring (e.g., chronic diseases such as diabetes). The vision of connected healthcare is growing because of the availability of new technological tools. By the application of the IoT and new technologies, it is possible to create a health application that appears every morning to request reading the level of glucose in the blood and collects data from the patient automatically [2]. In the vision of connected healthcare, patients are those who take control of their health and being in good physical and mental health due to this application. In addition, this leads to a good responsibility and control of heath by allowing a real scenario for the IoT in healthcare. IoT will help doctors to respond quickly in emergencies and allow them to cooperate with international hospitals to track the status of a patient. There are also other applications of IoT such as patient identification; this application aims to reduce adverse events for patients, maintenance of comprehensive electronic medical records [3, 4].

10.2 Big Data in Healthcare

Big data was defined as early as 2001. Doug Laney, an analyst at META (currently Gartner), has defined the challenges and opportunities generated by data growth with a 3Vs (that is, increasing volume, velocity, and variety [5]), 4Vs (we can add the veracity), and 5V models (we can also add the valeur) (Fig. 10.1).

Volume: refers to the huge amounts of data generated every second in the Healthcare field. Just think of all the emails, tweets, photos, videos, sensor data that we produce and share every second. We no longer speak Terabytes but Zettabytes or Brontobytes. Variety: refers to the different types of data we can use. In the past, we relied primarily on structured data. The type we can table and carefully organize, such as patient appointments, visits, etc. Less structured data, such as text files, analytic, photos, etc., were largely ignored. Today, we have the ability to use and analyze a wide variety of data, including written text, photos, ultrasound, as well as biometric data, and video content. Velocity: refers to the speed at which new

Fig. 10.1 Big data: the three V's

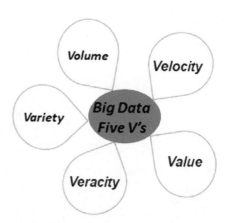

data is generated and moves. Just think of messages exchanged between doctors and also between patients who become viral in seconds, banking transactions done in minutes, or the time that software takes to analyze patient states and capture behaviors, must be milliseconds! Big data today allows us to analyze the data as it is generated, without having to analyze it in databases. The veracity and accuracy of data remains today the main challenge of big data. At present, these data are not yet sufficiently controlled, and the accuracy of the analyses is affected. The lack of veracity and quality of data costs about 3.1 trillion dollars a year in the United States. Value: it is the last V to consider when talking about big data. It is nice to have access to big data but still need to turn them into value; otherwise, it is useless! In this case, we can say that the value is a very important V!

10.2.1   Existing Technologies

The big data technologies involve commercial and open-source platforms and services for storage, security, access, and processing of data, many of them are based on the widely used open-source Hadoop framework. Hadoop is an independent Java-based programming framework that enhances the computation of large data sets in a distributed computing environment. Hadoop has two components [6] (Fig. 10.2):

- Hadoop distributed file system
- MapReduce

HDFS: HDFS is a distributed file system that provides high-performance access to data distributed in Hadoop clusters. MapReduce: MapReduce is one of the most adopted frameworks in the field of batch processing, and it is a programming model in which a MapReduce program can have two functions [6]: the map and the reduction, which requires moving data across the nodes. The map and the reduction, each defining a mapping of one set of key-value pairs to another. MapReduce is

Fig. 10.2 Components of
Hadoop

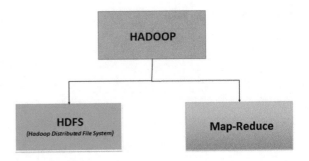

Table 10.1 Classification algorithms apply in Big Data

Algorithms	Application in Big Data
Decision Tree	Image classification, text categorization
Naïve Bayes	Text classification, sentiment analysis, opinion mining
K-NN	Face recognition, medical imaging data
SVM	Image classification, pattern recognition, hand-written recognition

based on the divide and conquer method, and works by recursively breaking down a complex problem into many sub-problems, until these sub-problems is scalable for solving directly [7].

10.2.2 Classification Data

Data classification is a process with many types of existing data sets for analysis, and the same type of data entities is extracted and divides the new and unknown data types according to the extracted data function [8]. There are six main classification models integrated in recent Weka tools, namely decision tree, ripper rule, neural networks, Naive Bayes, k-nearest neighbor, and support vector machine [7] (Table 10.1).

- Decision tree (J48) is a decision support system that uses a tree-shaped graph and its possible consequences. A decision tree, or a classification tree, is used to learn a classification function that concludes the value. A decision tree is a model that can predict the value of a target attribute (dependent variable) from the known values for a set of input attributes (independent variables) [9]. Decision trees are the most powerful approaches to knowledge discovery and data mining. It includes mass search technology and the complex mass of data to discover useful models. Decision trees are very effective tools in many areas such as data mining and text mining, information extraction, machine learning, and pattern recognition.
- The support vector machine (SVM) is a statistic-based learning method, which avoids the problems of difficult to determine, overlearning and underlearning,

and local minimum of network structure in the artificial neural networks [10, 11]. It shows many unique advantages in solving the classification and regression problems in terms of small sample, nonlinear and high dimension, which is considered as the best theory for the small sample classification, regression, and other issues [12].

- Ripper rule (JRIP) is used to generate various rules by adding repetitive data sets until the rules cover all data configurations according to the set of learning data. In addition, once all the rules are generated, some of them will be merged to reduce the size.
- Neural networks (MLPs—multilevel perceptron) have a distinctive feature as a three-layer feed-forward neural network: an input layer, a hidden layer, and an output layer. In order to link each node in each level, it may include additional weight to properly adjust the traversing path selection process.
- Naïve Bayes is derived from Bayes' theorem by applying the probabilistic learning knowledge for classification, assuming that the predictive attribute is conditionally independent according to each individual class. Naive Bayes classifier is a popular algorithm in machine learning. It is a supervised learning algorithm used for classification. It is particularly useful for text classification issues. An example of using Naive Bayes is that of the anti-spam filter.
- K-nearest neighbor (IBK) is used to perform the classification considering k subsets of data, each of them has similar characteristics by applying the Euclidean distance to understand the group, and here, IBK is the one of the k-nearest-simplified-neighbor classifiers.
- Vector support (sequential minimal optimization (SMO)) is basically a linear classifier (two classes) used to determine the largest distance between two sets, and SMO is the minimum sequential optimization algorithm for SVM training using polynomial or Gaussian kernels.

As shown in Fig. 10.3, the glucose sensor is used to sense the glucose values in the blood of the diabetic patient, and transfer the sensed data over short-range wireless communication to the patient's Android smartphone. The smartphone then aggregates and stores the sensed data, provides the healthcare monitoring interface to the patients for logging, and sends the physiological data to the medical server at a specified time interval whereby the physicians can directly have access for further analysis, diagnosis, and intervention. Today, the need for optimal management of the patient's health is very important and it becomes even more crucial if management is done remotely when the patient is at home [13]. That is why we tried to work on this system for diabetics.

Wearable sensors are the key components in the WBAN as they collect the vital data of the human body for further usage [14]. However, the blood glucose signs information can also be captured with the help of RFID technology and sensors through wearable devices [14]. A glucose concentration value ≤ 70 mg/dl is defined as hypoglycemic and a glucose concentration value ≥ 180 mg/dl is defined as hyperglycemic. If ever the threshold set is reached, an alarm is triggered whereby

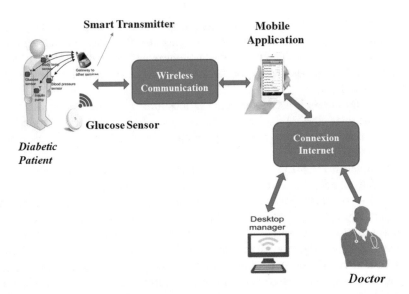

Fig. 10.3 Illustration of Blood Glucose measurement

Table 10.2 Glucose measurement

Day	Morning	Evening
1	1.34	0.78
2	2.34	0.87
3	1.11	1.02
4	1.45	0.55
5	0.88	1.43
6	1.00	0.78
7	0.99	1.43

the patient will receive a warning message on his mobile phone and similarly the physician will receive a warning message on the remote server.

10.3 Result and Discussion

In this section, we discuss the results of the testing system. The analysis is focused on the blood glucose data communication device as well as the performance of communication with Arduino GSM modem to send a short message service.

From the test results as shown in Table 10.2, it can be stated that the measurement data was successfully sent. When referring to the GSM communication the test results only showing text messages can show a waiting time difference or delay

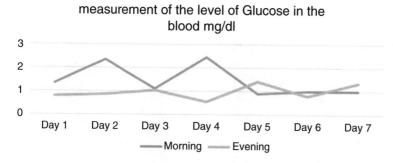

Fig. 10.4   Measurement of the level of Glucose in the blood

Table 10.3   The accuracy level

Algorithms	Correctly classified instances	Incorrectly classified instances
Naïve Bayes	91.8301%	8.1699%
RandomTree	99.6732%	0.3268%
ZeroR	69.6078%	30.3922%
J48	99.3464%	0.6536%

between sending and receiving SMS. Of 24 times of testing, the range of delay between transmitting and receiving SMS was from 33 to 122 s, and the average was 48.27 s. This result was related to the relatively slower data processing in the microcontroller as the microcontroller must acquire data from sensors in advance before being processed into a short message format and sent to the medical side. Figure 10.4 shows a graph that shows the measurement of the level of glucose in the blood.

The data set during this work is tested and analyzed with four classification algorithms those are Naïve Bayes, J48, ZeroR, and RandomTree (using training set). In addition, a comparison of accuracy of all classifiers is done and finally it has been investigated that RandomTree technique performs best with accuracy 99.6732 the accuracy level of all the algorithms are given in Table 10.3.

Table 10.3 summarizes performance measures for all classifiers; J48, ZeroR, RandomTree, and Naive Bayes. Figure 10.5 shows the comparative analysis of various classifiers in terms of correctly and incorrectly classified instances. The graph shows that the highest ranked is RandomTree, 99.67 and the lowest for ZeroR, 69.60. Figure 10.6 analyzes the study between accuracy and area under ROC and PRC. The accuracy is high for J48 with a value of 0.996 and low for ZeroR with a value of 0.696. The area under ROC is the highest for Naive Bayes with a value of 0.993 and low for the ZeroR with a value of 0.5. The area under PRC is the highest for RandomTree with a value of 0.993 and low for the ZeroR with a value of 0.696. Figure 10.7 shows the time graph of various classification algorithms. ZeroR takes the longest time with 0.05 s, RandomTree taking only 0.01 s.

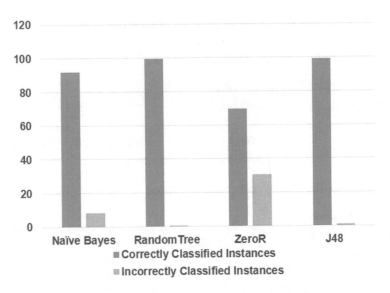

Fig. 10.5 The graph of correctly and incorrectly classified instances of algorithms

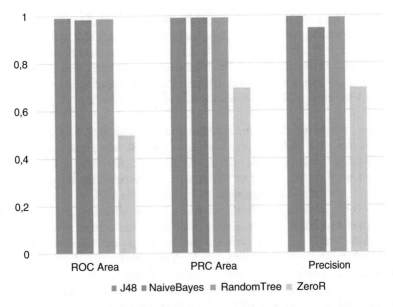

Fig. 10.6 Comparison of true classifiers based on area under ROC, PRC area, and precision

Fig. 10.7 The graph for training time results for top four classifiers

10.4   Conclusion

In this paper, we have discussed the issue of the classification of data in the healthcare field, especially in diabetic patients, our case study is based on one diabetic person, we measured the level of glucose for 30 days. We have raised the database of test measurements, we worked with four methods, the experimental analyses show that the Trees Random forest gives good results with very satisfactory detection rates. In perspective of this work we will perform the method of Trees Random forest and we will thus generalize both classes studied at once. The study compared the accuracy of the results obtained through four algorithms J48, ZeroR, Naive Bayes, and RandomTree. Even though the four algorithms were good enough in predicting the blood glucose using various parameters, RandomTree was found to be the best. It gave the most accurate result whether the patient had the possibility of blood glucose disease. This system can also be used in future projects to detect the specific type of blood glucose in particular. Thereby the diagnosis and management of blood glucose disease can be made simpler.

References

1. K.R. Ghani, K. Zheng, J.T. Wei, C.P. Friedman, European urology **66**(6), 975 (2014)
2. S. Sendra, L. Parra, J. Lloret, J. Tomás, Information Fusion **40**, 76 (2018)
3. T.R. Bennett, C. Savaglio, D. Lu, Massey, X. Wang, J. Wu, R. Jafari, Information Fusion **11**(1), 25 (2014). https://doi.org/10.1145/2633651.2637472. https://dl.acm.org/citation.cfm? id=2637472
4. H. Asri, H. Mousannif, H.A. Moatassime, T. Noel, 2015 International Conference on Cloud Technologies and Applications (CloudTech) **11**(1), 25 (2015). https://doi.org/10.1109/ CloudTech.2015.7337020. https://ieeexplore.ieee.org/document/7337020
5. A. Rghioui, S. Sendra, J. Lloret, A. Oumnad, Network Protocols and Algorithms **8**(3), 15 (2016)

6. W. Raghupathi, V. Raghupathi, Health information sciences and system (3), 25 (2014). https://doi.org/10.1186/2047-2501-2-3
7. C.P. Chen, C.Y. Zhang, Information sciences **275**, 314 (2014)
8. J.A. Stankovic, IEEE Internet of Things Journal **1**(1), 3 (2014). https://doi.org/10.1109/JIOT. 2014.2312291. http://ieeexplore.ieee.org/document/6774858/
9. T. Korting, Image Processing Division, National Institute for Space ... (2013). http://www. academia.edu/1983952/C4._5_algorithm_and_Multivariate_Decision_Trees
10. K. Avila, P. Sanmartin, D. Jabba, M. Jimeno, Sensors **17**(8), 1703 (2017). https://doi.org/10. 3390/s17081703. http://www.mdpi.com/1424-8220/17/8/1703
11. T. Wu, F. Wu, J.M. Redoute, M.R. Yuce, IEEE Access **5**, 11413 (2017). https://doi.org/10. 1109/ACCESS.2017.2716344. http://ieeexplore.ieee.org/document/7950903/
12. W. Yu, T. Liu, R. Valdez, M. Gwinn, M.J. Khoury, BMC Medical Informatics and Decision Making **10**(1), 16 (2010). https://doi.org/10.1186/1472-6947-10-16. http:// bmcmedinformdecismak.biomedcentral.com/articles/10.1186/1472-6947-10-16
13. M. Ali, S.C. Han, H.S.M. Bilal, S. Lee, M.J.Y. Kang, B.H. Kang, M.A. Razzaq, M.B. Amin, International Journal of Medical Informatics **109**, 55 (2018). https://doi.org/10.1016/j.ijmedinf. 2017.11.004. http://www.ncbi.nlm.nih.gov/pubmed/29195707https://linkinghub.elsevier.com/ retrieve/pii/S1386505617304161
14. V.N. Vapnik, *The Nature of Statistical Learning Theory* (Springer-Verlag, Berlin, Heidelberg, 1995)

Index

A
Adaptive control, 155
Aging, 56, 58, 61–63, 82, 83
Arithmetical, 73, 74
Artificial intelligence (AI), 58, 71, 83
   big data sciences, 58
   disciplines, 83
   education system, 142
   NDT 4.0., 71
   systemic NDT optimal design, 71
Asymptotic stability, 34–38

B
Backward stochastic differential equations
     (BSDE)
   convex polyhedron, 23–24
   notations, 22–23
   oblique reflection, 21
   penalization approach, 21
Bari's theorem, 48
Behavioral analysis system, 146
Big data
   algorithms, 167
   classification (*see* Data classification)
   classifiers, 167
   GSM communication, 166
   healthcare, 162
   SMS, 167
   technologies, 163–164
   3Vs
      vareity, 162
      velocity, 162
      volume, 162
      value, 163
Boundary conditions, 37, 38, 42, 47
Boundary control, 33
Boundary feedback, 33, 34
Brownian motion, 22

C
CAST3M Code
   FEM and XFEM, 124
   industrial calculation codes, 117
   operator Dall, 111
   structure calculation code, 124
Cauchy–Schwartz inequality, 14, 15
Chaos, 57, 70
Chirp-coded, 64, 65, 69, 70
Compact Tension (CT), 110, 126
Compute Unified Device Architecture
     (CUDA), 132
Connected vehicles
   cost and security, 158
   IoT applications, 155
   IP-based network protocol, 151
   ITS, 151
   road safety, 151
Conservative system, 49
Content-centric networking (CCN), 152
Content store (CS), 155–157
Convex polyhedral domain, 22
Correlation, 65
CPU time, 124, 125
Crack advance, 119, 123

© Springer Nature Switzerland AG 2020
S. Dos Santos et al. (eds.), *Recent Advances in Mathematics and Technology*,
Applied and Numerical Harmonic Analysis,
https://doi.org/10.1007/978-3-030-35202-8

Applied and Numerical Harmonic Analysis (97 volumes)

1. A. I. Saichev and W. A. Woyczyński: *Distributions in the Physical and Engineering Sciences* (ISBN: 978-0-8176-3924-2)
2. C. E. D'Attellis and E. M. Fernandez-Berdaguer: *Wavelet Theory and Harmonic Analysis in Applied Sciences* (ISBN: 978-0-8176-3953-2)
3. H. G. Feichtinger and T. Strohmer: *Gabor Analysis and Algorithms* (ISBN: 978-0-8176-3959-4)
4. R. Tolimieri and M. An: *Time-Frequency Representations* (ISBN: 978-0-8176-3918-1)
5. T. M. Peters and J. C. Williams: *The Fourier Transform in Biomedical Engineering* (ISBN: 978-0-8176-3941-9)
6. G. T. Herman: *Geometry of Digital Spaces* (ISBN: 978-0-8176-3897-9)
7. A. Teolis: *Computational Signal Processing with Wavelets* (ISBN: 978-0-8176-3909-9)
8. J. Ramanathan: *Methods of Applied Fourier Analysis* (ISBN: 978-0-8176-3963-1)
9. J. M. Cooper: *Introduction to Partial Differential Equations with MATLAB* (ISBN: 978-0-8176-3967-9)
10. Procházka, N. G. Kingsbury, P. J. Payner, and J. Uhlir: *Signal Analysis and Prediction* (ISBN: 978-0-8176-4042-2)
11. W. Bray and C. Stanojevic: *Analysis of Divergence* (ISBN: 978-1-4612-7467-4)
12. G. T. Herman and A. Kuba: *Discrete Tomography* (ISBN: 978-0-8176-4101-6)
13. K. Gröchenig: *Foundations of Time-Frequency Analysis* (ISBN: 978-0-8176-4022-4)
14. L. Debnath: *Wavelet Transforms and Time-Frequency Signal Analysis* (ISBN: 978-0-8176-4104-7)
15. J. J. Benedetto and P. J. S. G. Ferreira: *Modern Sampling Theory* (ISBN: 978-0-8176-4023-1)
16. D. F. Walnut: *An Introduction to Wavelet Analysis* (ISBN: 978-0-8176-3962-4)

© Springer Nature Switzerland AG 2020
S. Dos Santos et al. (eds.), *Recent Advances in Mathematics and Technology*,
Applied and Numerical Harmonic Analysis,
https://doi.org/10.1007/978-3-030-35202-8

17. A. Abbate, C. DeCusatis, and P. K. Das: *Wavelets and Subbands* (ISBN: 978-0-8176-4136-8)
18. O. Bratteli, P. Jorgensen, and B. Treadway: *Wavelets Through a Looking Glass* (ISBN: 978-0-8176-4280-80)
19. H. G. Feichtinger and T. Strohmer: *Advances in Gabor Analysis* (ISBN: 978-0-8176-4239-6)
20. O. Christensen: *An Introduction to Frames and Riesz Bases* (ISBN: 978-0-8176-4295-2)
21. L. Debnath: *Wavelets and Signal Processing* (ISBN: 978-0-8176-4235-8)
22. G. Bi and Y. Zeng: *Transforms and Fast Algorithms for Signal Analysis and Representations* (ISBN: 978-0-8176-4279-2)
23. J. H. Davis: *Methods of Applied Mathematics with a MATLAB Overview* (ISBN: 978-0-8176-4331-7)
24. J. J. Benedetto and A. I. Zayed: *Sampling, Wavelets, and Tomography* (ISBN: 978-0-8176-4304-1)
25. E. Prestini: *The Evolution of Applied Harmonic Analysis* (ISBN: 978-0-8176-4125-2)
26. L. Brandolini, L. Colzani, A. Iosevich, and G. Travaglini: *Fourier Analysis and Convexity* (ISBN: 978-0-8176-3263-2)
27. W. Freeden and V. Michel: *Multiscale Potential Theory* (ISBN: 978-0-8176-4105-4)
28. O. Christensen and K. L. Christensen: *Approximation Theory* (ISBN: 978-0-8176-3600-5)
29. O. Calin and D.-C. Chang: *Geometric Mechanics on Riemannian Manifolds* (ISBN: 978-0-8176-4354-6)
30. J. A. Hogan: *Time?Frequency and Time?Scale Methods* (ISBN: 978-0-8176-4276-1)
31. C. Heil: *Harmonic Analysis and Applications* (ISBN: 978-0-8176-3778-1)
32. K. Borre, D. M. Akos, N. Bertelsen, P. Rinder, and S. H. Jensen: *A Software-Defined GPS and Galileo Receiver* (ISBN: 978-0-8176-4390-4)
33. T. Qian, M. I. Vai, and Y. Xu: *Wavelet Analysis and Applications* (ISBN: 978-3-7643-7777-9)
34. G. T. Herman and A. Kuba: *Advances in Discrete Tomography and Its Applications* (ISBN: 978-0-8176-3614-2)
35. M. C. Fu, R. A. Jarrow, J.-Y. Yen, and R. J. Elliott: *Advances in Mathematical Finance* (ISBN: 978-0-8176-4544-1)
36. O. Christensen: *Frames and Bases* (ISBN: 978-0-8176-4677-6)
37. P. E. T. Jorgensen, J. D. Merrill, and J. A. Packer: *Representations, Wavelets, and Frames* (ISBN: 978-0-8176-4682-0)
38. M. An, A. K. Brodzik, and R. Tolimieri: *Ideal Sequence Design in Time-Frequency Space* (ISBN: 978-0-8176-4737-7)
39. S. G. Krantz: *Explorations in Harmonic Analysis* (ISBN: 978-0-8176-4668-4)
40. B. Luong: *Fourier Analysis on Finite Abelian Groups* (ISBN: 978-0-8176-4915-9)

41. G. S. Chirikjian: *Stochastic Models, Information Theory, and Lie Groups, Volume 1* (ISBN: 978-0-8176-4802-2)
42. C. Cabrelli and J. L. Torrea: *Recent Developments in Real and Harmonic Analysis* (ISBN: 978-0-8176-4531-1)
43. M. V. Wickerhauser: *Mathematics for Multimedia* (ISBN: 978-0-8176-4879-4)
44. B. Forster, P. Massopust, O. Christensen, K. Gröchenig, D. Labate, P. Vandergheynst, G. Weiss, and Y. Wiaux: *Four Short Courses on Harmonic Analysis* (ISBN: 978-0-8176-4890-9)
45. O. Christensen: *Functions, Spaces, and Expansions* (ISBN: 978-0-8176-4979-1)
46. J. Barral and S. Seuret: *Recent Developments in Fractals and Related Fields* (ISBN: 978-0-8176-4887-9)
47. O. Calin, D.-C. Chang, and K. Furutani, and C. Iwasaki: *Heat Kernels for Elliptic and Sub-elliptic Operators* (ISBN: 978-0-8176-4994-4)
48. C. Heil: *A Basis Theory Primer* (ISBN: 978-0-8176-4686-8)
49. J. R. Klauder: *A Modern Approach to Functional Integration* (ISBN: 978-0-8176-4790-2)
50. J. Cohen and A. I. Zayed: *Wavelets and Multiscale Analysis* (ISBN: 978-0-8176-8094-7)
51. D. Joyner and J.-L. Kim: *Selected Unsolved Problems in Coding Theory* (ISBN: 978-0-8176-8255-2)
52. G. S. Chirikjian: *Stochastic Models, Information Theory, and Lie Groups, Volume 2* (ISBN: 978-0-8176-4943-2)
53. J. A. Hogan and J. D. Lakey: *Duration and Bandwidth Limiting* (ISBN: 978-0-8176-8306-1)
54. G. Kutyniok and D. Labate: *Shearlets* (ISBN: 978-0-8176-8315-3)
55. P. G. Casazza and P. Kutyniok: *Finite Frames* (ISBN: 978-0-8176-8372-6)
56. V. Michel: *Lectures on Constructive Approximation* (ISBN : 978-0-8176-8402-0)
57. D. Mitrea, I. Mitrea, M. Mitrea, and S. Monniaux: *Groupoid Metrization Theory* (ISBN: 978-0-8176-8396-2)
58. T. D. Andrews, R. Balan, J. J. Benedetto, W. Czaja, and K. A. Okoudjou: *Excursions in Harmonic Analysis, Volume 1* (ISBN: 978-0-8176-8375-7)
59. T. D. Andrews, R. Balan, J. J. Benedetto, W. Czaja, and K. A. Okoudjou: *Excursions in Harmonic Analysis, Volume 2* (ISBN: 978-0-8176-8378-8)
60. D. V. Cruz-Uribe and A. Fiorenza: *Variable Lebesgue Spaces* (ISBN: 978-3-0348-0547-6)
61. W. Freeden and M. Gutting: *Special Functions of Mathematical (Geo-)Physics* (ISBN: 978-3-0348-0562-9)
62. A. I. Saichev and W. A. Woyczyñski: *Distributions in the Physical and Engineering Sciences, Volume 2: Linear and Nonlinear Dynamics of Continuous Media* (ISBN: 978-0-8176-3942-6)
63. S. Foucart and H. Rauhut: *A Mathematical Introduction to Compressive Sensing* (ISBN: 978-0-8176-4947-0)

64. G. T. Herman and J. Frank: *Computational Methods for Three-Dimensional Microscopy Reconstruction* (ISBN: 978-1-4614-9520-8)
65. A. Paprotny and M. Thess: *Realtime Data Mining: Self-Learning Techniques for Recommendation Engines* (ISBN: 978-3-319-01320-6)
66. A. I. Zayed and G. Schmeisser: *New Perspectives on Approximation and Sampling Theory: Festschrift in Honor of Paul Butzer's 85th Birthday* (ISBN: 978-3-319-08800-6)
67. R. Balan, M. Begue, J. Benedetto, W. Czaja, and K. A. Okoudjou: *Excursions in Harmonic Analysis, Volume 3* (ISBN: 978-3-319-13229-7)
68. H. Boche, R. Calderbank, G. Kutyniok, and J. Vybiral: *Compressed Sensing and its Applications* (ISBN: 978-3-319-16041-2)
69. S. Dahlke, F. De Mari, P. Grohs, and D. Labate: *Harmonic and Applied Analysis: From Groups to Signals* (ISBN: 978-3-319-18862-1)
70. A. Aldroubi: *New Trends in Applied Harmonic Analysis* (ISBN: 978-3-319-27871-1)
71. M. Ruzhansky: *Methods of Fourier Analysis and Approximation Theory* (ISBN: 978-3-319-27465-2)
72. G. Pfander: *Sampling Theory, a Renaissance* (ISBN: 978-3-319-19748-7)
73. R. Balan, M. Begue, J. Benedetto, W. Czaja, and K. A. Okoudjou: *Excursions in Harmonic Analysis, Volume 4* (ISBN: 978-3-319-20187-0)
74. O. Christensen: *An Introduction to Frames and Riesz Bases, Second Edition* (ISBN: 978-3-319-25611-5)
75. E. Prestini: *The Evolution of Applied Harmonic Analysis: Models of the Real World, Second Edition* (ISBN: 978-1-4899-7987-2)
76. J. H. Davis: *Methods of Applied Mathematics with a Software Overview, Second Edition* (ISBN: 978-3-319-43369-1)
77. M. Gilman, E. M. Smith, and S. M. Tsynkov: *Transionospheric Synthetic Aperture Imaging* (ISBN: 978-3-319-52125-1)
78. S. Chanillo, B. Franchi, G. Lu, C. Perez, and E. T. Sawyer: *Harmonic Analysis, Partial Differential Equations and Applications* (ISBN: 978-3-319-52741-3)
79. R. Balan, J. Benedetto, W. Czaja, M. Dellatorre, and K. A. Okoudjou: *Excursions in Harmonic Analysis, Volume 5* (ISBN: 978-3-319-54710-7)
80. I. Pesenson, Q. T. Le Gia, A. Mayeli, H. Mhaskar, and D. X. Zhou: *Frames and Other Bases in Abstract and Function Spaces: Novel Methods in Harmonic Analysis, Volume 1* (ISBN: 978-3-319-55549-2)
81. I. Pesenson, Q. T. Le Gia, A. Mayeli, H. Mhaskar, and D. X. Zhou: *Recent Applications of Harmonic Analysis to Function Spaces, Differential Equations, and Data Science: Novel Methods in Harmonic Analysis, Volume 2* (ISBN: 978-3-319-55555-3)
82. F. Weisz: *Convergence and Summability of Fourier Transforms and Hardy Spaces* (ISBN: 978-3-319-56813-3)
83. C. Heil: *Metrics, Norms, Inner Products, and Operator Theory* (ISBN: 978-3-319-65321-1)
84. S. Waldron: *An Introduction to Finite Tight Frames: Theory and Applications.* (ISBN: 978-0-8176-4814-5)

85. D. Joyner and C. G. Melles: *Adventures in Graph Theory: A Bridge to Advanced Mathematics*. (ISBN: 978-3-319-68381-2)
86. B. Han: *Framelets and Wavelets: Algorithms, Analysis, and Applications* (ISBN: 978-3-319-68529-8)
87. H. Boche, G. Caire, R. Calderbank, M. März, G. Kutyniok, and R. Mathar: *Compressed Sensing and Its Applications* (ISBN: 978-3-319-69801-4)
88. A. I. Saichev and W. A. Woyczyñski: *Distributions in the Physical and Engineering Sciences, Volume 3: Random and Fractal Signals and Fields* (ISBN: 978-3-319-92584-4)
89. G. Plonka, D. Potts, G. Steidl, and M. Tasche: *Numerical Fourier Analysis* (978-3-030-04305-6)
90. K. Bredies and D. Lorenz: *Mathematical Image Processing* (ISBN: 978-3-030-01457-5)
91. H. G. Feichtinger, P. Boggiatto, E. Cordero, M. de Gosson, F. Nicola, A. Oliaro, and A. Tabacco: *Landscapes of Time-Frequency Analysis* (ISBN: 978-3-030-05209-6)
92. E. Liflyand: *Functions of Bounded Variation and Their Fourier Transforms* (978-3-030-04428-2)
93. R. Campos: *The XFT Quadrature in Discrete Fourier Analysis* (978-3-030-13422-8)
94. M. Abell, E. Iacob, A. Stokolos, S. Taylor, S. Tikhonov, J. Zhu: *Topics in Classical and Modern Analysis: In Memory of Yingkang Hu* (978-3-030-12276-8)
95. H. Boche, G. Caire, R. Calderbank, G. Kutyniok, R. Mathar, P. Petersen: *Compressed Sensing and its Applications: Third International MATHEON Conference 2017* (978-3-319-73073-8)
96. A. Aldroubi, C. Cabrelli, S. Jaffard, U. Molter: *New Trends in Applied Harmonic Analysis, Volume II: Harmonic Analysis, Geometric Measure Theory, and Applications* (978-3-030-32352-3)
97. S. Dos Santos, M. Maslouhi, K. Okoudjou: *Recent Advances in Mathematics and Technology: Proceedings of the First International Conference on Technology, Engineering, and Mathematics, Kenitra, Morocco, March 26-27, 2018* (978-3-030-35201-1)

For an up-to-date list of ANHA titles, please visit http://www.springer.com/series/4968

Printed in the United States
By Bookmasters